Conceptions of Space and Time

Conceptions of Space and Time
Sources, Evolution, Directions

Murad D. Akhundov
translated by Charles Rougle

The MIT Press
Cambridge, Massachusetts
London, England

English translation © 1986 by the Massachusetts Institute of Technology

This work was originally published in Russian as *Kontseptsii Prostranstva i Vremeni: Istoki, Evolyutsiya, Perspektivy*, © 1982 by Nauka Publishing House, Moscow, USSR.

All rights reserved. No part of this book may be reproduced in any form or by any electronic or mechanical means (including photocopying, recording, or information storage and retrieval) without permission in writing from the publisher.

This book was set in Baskerville by Achorn Graphic Services and was printed and bound by The Murray Printing Co. in the United States of America.

Library of Congress Cataloging in Publication Data

Akhundov, Murad Davudovich.
  Conceptions of space and time—sources, evolution, directions.

  Translation of: Konseptsii prostranstva i vremeni—istoki, evoliutsiia, perspektivy.
  Bibliography: p.
  Includes index.
  1. Space and time—History. I. Title.
BD632.A4513 1986   115   86-8777
ISBN 0-262-01091-7

Contents

Introduction *1*

1
Sources of Spatial and Temporal Conceptions *11*

Human Evolution and the Origin of Spatial and Temporal Notions *13*

The Evolution of Spatial and Temporal Conceptions in the Early Stages of Human Culture *30*

2
The Philosophical Evolution of Spatial and Temporal Conceptions *55*

Space and Time in Early Greek Philosophy *55*

The Evolution of Conceptions of Space and Time in Medieval and Modern Philosophy *88*

3
A Philosophical Analysis of Space and Time in Physical Theory *125*

The Status of Space and Time in the Early History of Classical Mechanics *125*

Status and Directions of Spatial and Temporal Conceptions in Modern Physics: Theoretical and Empirical Structures. The Possible Macroscopic Nature of Space and Time *138*

Space and Time in the Axiomatic Approach to Physical Theory *152*

Notes *167*
Index *195*

Conceptions of Space and Time

Introduction

Among the most fundamental notions of culture, conceptions of space and time have figured prominently in the history of human thought. These conceptions are so basic that, at one stage of cultural development, in the ancient mythological, religious, and philosophical systems, they were regarded as the source of the world (*Zurvan* of early Zoroastrianism, Chaos in ancient Greek mythology, *akasha* and *kala* in Indian philosophy). In subsequent cultural history, space and time continued to be bound up with the foundations of the universe; in early Greek materialist doctrine they are one of the sources of the world (Democritus's void); in objective idealist thought, space is the *sensorium Dei;* in Kantian subjective idealist systems, space and time are elevated to the rank of a priori forms of intuition; finally, dialectical materialism regards them as universal forms of existence of matter.

Spatial and temporal notions also occupy a vital position in modern science. In physics, for example, which has become the vanguard of the natural sciences, space and time are regarded as basic concepts, because most physical notions are introduced by means of operational rules employing spatial and/or temporal distances. This indicates that space and time are empirically fundamental to physics, but, particularly if one adheres to the geometrodynamic doctrine, which holds that nothing exists in the universe but empty curved space, it can easily be established that they are to some extent theoretically basic as well.

Such spatial theories of matter indicate that, each time science enters a new loop in the spiral of knowledge, it has recourse to concepts that are consonant with notions held by our distant ancestors, who came close to "geometrodynamic" ideas in their mythology. It almost seems as though the farther science advances, the deeper we must probe for the precursors of modern concepts. Today we can see that the level of the basic structure of matter to which science has penetrated has a parallel in "insane ideas" deriving not only from early Greek natural philosophy (generally regarded as inexhaustibly generating the prototypes of contemporary notions) but also from even deeper strata in human cultural and intellectual achievement.

The process of rejection and revision of scholastic notions, by which science developed, frequently brings us back to the Greek sages. Today we have reached a boundary at which the "insane ideas" involve a rejection of basic notions common to early Greek natural philosophy, Scholasticism, classical science, and even the early stages of nonclassical natural science. Indeed, besides refuting the apriorism of physical geometry, modern physics is being obliged to revise the logic of physical theories, which are now developing in a non-Aristotelian context. Forced to take into account such ideas as relative cardinality, tolerance, and fuzzy sets, present-day science is in the process of revising the concept of the classical continuum.

Lenin once pointed out the connection between fantasy and the seeds of scientific thought, noting the varying interrelationship of science and mythology at different stages of development. He concludes: "But now! the same thing, the same connection, but the ratio between science and myth has changed."[1] It is thus understandable that Hermann Bondi[2] should show an interest in assumptions and myths in physical theory, that Karl Pribram[3] should argue for the existence of what he calls neuromythology, and that Hannes Alfven[4] should reflect on whether modern cosmology is a myth or a science. We can also appreciate the difficulties encountered by Western scientific methodology in formulating criteria for distinguishing among science, philosophy, and mythology. Either science is successfully separated from philosophy, in which case the latter becomes indistinguishable from mythology, or science itself merges with mythology.

I agree that the role of myths in scientific theories deserves to be taken into consideration. It is no less important, however, to try to discover mythological archetypes that may be of use to modern science and are perhaps even consonant with future scientific concepts.[5] These archetypes do not disappear without a trace in cultural history; they survive latently in diverse intellectual constructs of subsequent ages and cultures and are revived when circumstances become propitious at some new and higher spire of knowledge.

The notion of the continuum in classical mathematics and the atomic theory of classical physics are organically related to Anaxagoras's concept of infinite divisibility and Democritus's atomism, but the prototypes of, say, Lotfi Zadeh's fuzzy sets or the classes of N. Ya. Vilenkin and Yu. Shreider[6] are to be found in primitive notions of "a fluid multiplicity of vaguely defined elements," as discussed by Lévy-Bruhl; the concept of the bootstrap in modern subatomic physics[7] is in keeping with the premises of Zen, and the spatial model of matter also has its mythological precursors.

Many Western cultural figures, such as J. D. Salinger, Henri Matisse, Gustav Mahler, Martin Heidegger, and Albert Schweitzer, have attempted to overcome traditional Eurocentrism. Nor has the general influence of ancient Indian and oriental philosophy been limited exclusively to the humanities in the West. Although ancient Indian philosophy is customarily considered consonant with, say, existentialism, owing to the antipositivist and antiscientist attitude of existentialism, this in no way implies that natural science has not drawn on the rich conceptual heritage of Indian philosophy in its development. And the point here is not merely that exotic analogies can be established, for example, between Murray Gell-Mann's system of elementary particles and the Eightfold Path of the ancient Hindus or between Geoffrey Chew's bootstrap model and the parable of Indra's net of pearls, but that modern physics has devoted a growing interest to mythological and ancient oriental conceptions as it revises basic and tried notions, such as elementariness,[8] in its quest for new and revolutionary ideas. Modern natural science is interested not only in colorful myths but also in the structural aspects of the mythological world and the archaic cosmological constructs that have generated our notions of time and space.

Einstein once noted that "scientific thought is a development of prescientific thought. As the concept of space was already fundamental in the latter, we should begin with the concept of space in prescientific thought."[9] I think that investigating the sources, history, present status, and probable future directions of spatial and temporal notions is a task of great urgency. Such an inquiry need not begin with ancient natural philosophy, as most histories of these concepts do,[10] because the pre-Socratic philosophy of nature is often apprehended through the ideational system of Aristotle. Certain studies may be restricted to such an approach, although it must be borne in mind that "such specifically Peripatetic notions as ὕλη (matter), δτοιχετον (elements), and τό ὑποκείμενον (substance) are not satisfactory tools for the description of sixth-century ideas."[11] Referring to the sources of a particular concept becomes even more difficult when early Greek philosophy is merely an intermediate link between animistic, magical, and mythological notions on the one hand and scientific concepts of the modern age on the other.

The spatial and temporal conceptions of Greek natural philosophy themselves derive from earlier cultural and intellectual strata. Natural philosophy, therefore, must be analyzed not on the basis of its "final product," which to some extent the Aristotelian system may be considered to be, but rather with reference to its mythological foundations. Moreover, if we want to penetrate to the sources of spatial and temporal notions, not only do we have to turn to these Greek mythological roots, but we also must consider the specific features of primitive thought, examining (or at least charting) the origin of these concepts in mankind's biological and psychological evolution.

It should be emphasized that these problems are not only narrowly pertinent at the level of, say, developmental psychology but are also more broadly relevant within the wider context of logical and methodological studies on the structure of scientific knowledge. Thus, especially in connection with advances in quantum physics, many scholars have come to the conclusion that the separation of physics and psychology in modern science represents a serious shortcoming that can be remedied only by developing a more integrated view of the totality of our experience.

Such integrated psychological and physical approaches appear promising. Human perception is in a certain sense capable of "outstripping" physics. For example, it is generally assumed that mankind's distinctive features, in particular its macroscopic nature, account for the fact that the observational structures of physical theories are a macroworld structure, by which, in effect, is meant the time and space of classical mechanics. There is even a connection between this proposition and Bohr's well-known epistemological principle, which states that all experimental facts (in explanations of nonclassical phenomena) should be described in terms of classical concepts. This statement subtly changes into another one that at first glance seems fairly obvious, namely that mankind's specifically human features (perception and thought) correspond to the classical notions of space and time underlying all experience. This thesis may also be regarded as an epistemological principle, and it is a rather productive one in physics because it holds that classical space and time are epistemologically universal.[12] We must bear in mind, however, that, first, the thesis cannot be reduced to that of Bohr and, second, it is not a necessary principle. I examine this thesis in detail but restrict myself for the moment to noting that there are weighty reasons for the view that, essentially and with respect to general laws, our perception of the world is far closer to the spirit of relativistic physics than to that of classical physics (it must be remembered that the construction of the former signified a revolution in twentieth-century physics). David Bohm has devoted an interesting study to this important question of adequately understanding the applicability of everyday experience.[13]

Here, however, we are confronted by the startling fact that perception in the spirit of relativistic physics is not so much qualitatively new as it is a completely forgotten or uncognizable perception of the world. Indeed, spatiotemporal awareness in primitive societies is much in keeping with the relativistic spirit (V. G. Bogoraz (Tan)). The perception of the child is also closer to Einstein than to Newton. In their phylogenetic and ontogenetic childhood, humans perceive the world relativistically, and this, of course, demands that we scrutinize the specific character of primitive and infantile spatial and temporal perception. Moreover, even adult perception is not completely subject to the canons of classical mechanics, and our visual space is non-Euclidean. For two thousand years now, mankind has projected its kinesthetic space onto the world and elevated it to the rank of apriority and universality. It was the physicalization of this space that led to the formulation of classical mechanics.

My purpose in citing these facts here is not only to establish that we must refer to the sources of spatial and temporal concepts but also to demonstrate that psychological, logicohistorical, physical, and other problems are organically interrelated in the study of these notions. Physical theory is often treated as an immediate development and extension of our perception. Indeed, perception is abstract, and evolving physical theories serve as eyeglasses that enable us to view the world in increasingly sharp and comprehensive detail.[14] Let us not forget, however, that in certain respects our perception may correspond not to contemporary theories but to future physical ones. The world was perceived long before it was theoretically known. Naturally, perception carries a certain theoretical (mystical, mythological, etc.) load, but it also bears an impression of the real world (not merely an impression of a theoretical impression), meaning objective laws and features that may not be reflected yet in science and its theories.

In the course of their long evolutionary history, humans have achieved increasingly optimal adaptation to their surroundings. This means in turn that the structure of the world has more and more adequately stamped itself on mankind. It is interesting to note that modern cosmologists (P. Dicke, B. Carter, S. Hawking, and others) operate with a special anthropic principle that asserts that the present order of the universe is dependent on the fact of mankind's existence. That is, the structure of the universe is determined by a certain biological selection of fundamental physical constants.[15] Because mankind can exist only in a certain universe, observing humanity can teach us a great deal about that universe. Thus a study of how we perceive the world can greatly enrich our knowledge of that world and of our notions about ways and methods of scientifically investigating it.[16]

Here I think we can glimpse a rather fruitful approach to an analysis of the various primitive, infantile, mythological, and religious sources of spatial and temporal notions. This is the basis of the present inquiry into such conceptions in early Greek philosophy; it is to be hoped that such an approach will, on the one hand, spare us the feeling of scholars such as John Burnet[17] that this majestic intellectual edifice arose on a barren spot rather like Athena springing from the head of Zeus. (This, of course, does not imply that in a burst of atheistic enthusiasm we should think that Zeus's head is empty and view Athena's birth in the spirit of John Wheeler.) On the other hand, I will also avoid the opposite extreme, as represented by Heinrich Ritter's description of Pythagoras: "It is said that Pythagoras's teachers were the following: of geometry—the Egyptians; of arithmetic—the Phoenicians; of astronomy—the Chaldeans; of religion and morals—the Magi. This leaves nothing for the Greeks, so that Pythagoras may be considered an oriental sage."[18]

Early Greek philosophy, of course, is inconceivable apart from the mythological foundation that it not only remolded, reconstructed, rationalized, or superseded but also assimilated. The Greek sense of history, for example, depended for its development on the complete resurrection of mythological time, which was accomplished in a reflective philosophical form by Plato and Aristotle. It was this that A. V. Losev had in mind when he noted that "the entire world of antiquity is permeated by mythology."[19] Certain pictures illustrating the history of Greek science are symbolic in this sense. Thus, on the cover of Marshall Clagget's book *Greek Science in Antiquity*, we find a cracked egg situated on a field of geometric lines continuing on to its shell, and from it is emerging, not a chick but a full-grown, bearded Greek philosopher.[20] Here we have an example of dialectical symbolism in the spirit of Jerzy Lec's famous aphorism: "Reaching the bottom, I heard a knock from below."

In my view, the distinctiveness of early Greek natural philosophy (including that of such leading representatives as Pythagoras) is in no way diminished by the organic bond and continuity that link it with mythology and ancient oriental wisdom (the *Enuma elish*, Hesiod's *Theogony*, etc.). Greek philosophy is a special stage of development in a world view that possesses its own mode of thought, its own language, epistemology, and methodology, and—a point of special interest to us—its own conceptions of space and time. Consequently, we must analyze its entire spectrum of substantial, attributive (which in turn breaks down into extensional and relational), static, continual, and atomic notions and then trace their career in the Christian world view, which refracted Greek wisdom through the prism of biblical concepts rooted in ancient oriental culture. Here certain conceptions were transformed through their participation in the Divine, whereas others were theologically eliminated or, as it were, "anathemized."

The Christian world view and its ordering of space and time present considerable interest, as they encompass the paradigm of an enormous historical period of human culture.[21] It was within this paradigm and through struggle with it that the rationalistic philosophy and science of the modern age were born and developed. The revival of many Greek doctrines and their corresponding spatial and temporal conceptions is organically related to these processes. Moreover, as we today examine the development of modern philosophy and science, we can fairly often discern the Greek notions in the clashes that occurred between different systems and currents. This applies to Descartes's extensional concept, Leibniz's relational idea, and Kant's subjective a priori conception. The classical concepts, of course, were not merely borrowed from the early Greeks but underwent significant changes as they were incorporated into different systems of philosophy and natural science and came to function in different sociocultural environments. The Democritian void, for example, was substantially transformed and enriched when it emerged as Newtonian empty space. If Democritus's void is merely a necessary condition for the movement of atoms, Newton's exhibits a wide variety of hypostases and defines the basic laws and principles of symmetry of classical mechanics.

Because a logicohistorical study cannot embrace all aspects of the evolution of spatial and temporal conceptions as a unified process, I am obliged to resort to various reconstructions. Development from the early Greeks to Newton, for example, cannot be considered only on the level of philosophical concepts. It must also be borne in mind that antiquity had its own mechanics developed primarily on an informal level. Thus Aristotle's mechanics operated in a corresponding space-time and has a structure that can be reconstructed fairly clearly,[22] and this provides a basis for a reconstructional approach to the structure of space-time in the history of mechanics.[23] Another method of reconstruction has its sources in ancient Greek geometry as presented axiomatically in Euclid's *Elements*.

Thus three types of reconstruction can be used to investigate the evolution of spatial and temporal conceptions: (1) conceptual evolution in the history of philosophy, (2) evolution as a successive broadening of geometrical axiomatics along the line "geometry → kinetics → dynamics," (3) the evolution of the structure of space-time through rival and successive mechanical systems (Aristotelian, Cartesian, Newtonian, etc.). Such a comprehensive logicohistorical and historicophilosophical analysis will enable us to investigate productively the status of space and time in classical physics (in Newton's *Mathematical Principles of Natural Philosophy,* which was in fact the first physical theory in the modern sense of the word) and in various logicomathematical reconstructions of this system (Euler, Lagrange, Hamilton, and others).

Newton's physics was so brilliantly successful that its universality was gradually absolutized in notions that reached their most extreme expression in mechanism. Yet Newton's theoretical system also had its weak points. Some difficulties were generated by optical phenomena that, when analyzed, led to the triumph of the rival wave theory of Huygens, Young, and Fresnel. Although the wave theory of light offered an explanation of optical problems, such as interference and diffraction, it gave rise to more general difficulties in the mechanistic world view.

For, in fact, this theory upset the view that everything real can be conceived as the motion of particles in space. Light waves were, after all, nothing more than undulatory states of empty space, and space thus gave up its passive role as a mere stage for physical events.[24]

Nor was the mechanistic Weltanschauung rescued by the numerous palliatives that were proposed. Space came to life, as followed even more clearly from the electrodynamics of Maxwell and Lorentz. "The physical states of space itself were the final physical reality."[25]

Nineteenth-century scientific advances irresistibly led naturalists toward a refutation of received spatial and temporal conceptions. Field theory proved incompatible with the instantaneous action-at-a-distance, on which the absolute simultaneity of events of Newton's world was based. There was also (positivist) criticism of absolute space and time on the philosophical level. A complicated state of affairs arose in physics because, although this heightened criticism had touched on fundamental questions, it could not contribute constructive solutions to the problems and paradoxes that arose. And then, of course, there are the sundry idealistic and spiritualistic schools and pseudoschools that thrive like parasites on the difficulties of natural science; the crisis physics went through at the turn of the century was no exception.

A number of substantial achievements in physics and philosophy went far toward clearing up this situation. Deserving note here are, first, Einstein's theory of relativity and the development of quantum mechanics by Bohr and Planck, and, second,

Lenin's consistent Marxist analysis of the impasse in *Materialism and Empirocriticism*. This analysis provides a reliable basis for the study of spatial and temporal notions in modern physics.

In what follows I consider the essence of the transition from classical to relativistic physics and the changes in spatial and temporal conceptions that accompanied that transition. It is rather widely held that this process involved a substitution of relational notions for substantial ones. In fact, however, the matter is not so simple, and in this study I attempt a more adequate analysis based on a consideration of observational and theoretical levels in the structure of different physical theories.

An even more complex state of affairs arose as physics moved toward quantum mechanics and the theory of elementary particles. Suffice it to mention the simple fact that the thesis on the macroscopic nature of space and time enjoys a certain popularity in the West (E. Zimmerman, G. Chew, and others). Such suspicions that space and time are nonuniversal, of course, demand thorough analysis, and here we will find it helpful to base ourselves on the same consideration of the observational and theoretical levels of physical theory.

In conclusion I examine various approaches to the axiomatization of physical theory. Specifically, I provide a critical analysis of Mario Bunge's objections to the status of space and time in axiomatization and, on the basis of familiar notions advanced by the Soviet physicists M. I. Podgoretskiy and Ya. A. Smorodinskiy, present the idea of a hierarchical axiomatics defined with the help of a metatheoretical regulator. The notion of a metatheoretical regulator for the construction of systems of physical knowledge can shed new light on the unity of physics. This principle is not reducible to attempts to universalize specific physical positions but presupposes that a metatheoretical regulator can be introduced to define a scheme for constructing a hierarchical axiomatics of physical systems, a program for generating its various conceptual structures. By this is meant not a search for a single universalized space but a program for generating the various geometries or spaces that correspond to different physical theories.

1

Sources of Spatial and Temporal Conceptions

I begin my discussion of the sources of spatial and temporal notions with an elegant and correct observation by S. I. Vavilov, the prominent Soviet physicist and expert on the philosophical problems of natural science:

> Side by side with the discoveries and insights of modern natural science, we find that the conceptual worlds of the child and primitive man and the world of the poet intentionally or unintentionally emulating them still exist. It is at times worthwhile to peek into them, for they are one possible source of scientific hypotheses. In these amazing, fantastic worlds bridges are boldly thrown between natural phenomena, establishing connections which science has sometimes never even suspected. Some of these links are guessed correctly, others are fundamentally wrong and simply absurd, but all of them deserve attention, as such mistakes often help us understand the truth.[1]

Unfortunately, modern studies devoted to logic, the methodology and history of science, and the philosophical problems of the natural sciences seldom address the origin of such basic notions as space and time. The attempts that have been made to analyze the problem historically generally begin with the Copernican revolution or, at best, with a cursory excursion into the modernized doctrines of early Greek natural philosophy and seldom give much consideration to the psychological and mythological aspects of the question. One rare exception is the English cosmologist G. J. Whitrow's *The Natural Philosophy of Time*, in which time as treated by physics and mathematics is preceded by a chapter on individual time. There he mainly discusses psychological and neurophysical aspects, such as biological time, the psychology and physiology of memory, and the personal identity, although he also touches on the psychological and social evolution of the idea of time.

The problem Whitrow addresses is of considerable contemporary interest, but it must also be observed that most of the materials he presents are quite outdated. Recently, interesting data have been obtained from both developmental psychology and studies of space and time in mythology, sagas, ethnogeny, and pathology. Whitrow fails to take into account the valuable contributions Soviet scholars have made in these areas, and his study suffers from the use of certain dubious notions. He maintains, for example, that the mind is a process that has

temporal extension and temporal position, but no spatial extension nor precise spatial location, although it has a field of influence which is strongest in the region occupied by the particular brain with which it is normally associated. This field of influence may, however, on occasion extend far beyond, as is clear from the now generally accepted evidence of telepathy.[2]

Such statements contribute little to a study of the origins of spatial and temporal conceptions. We are interested not in the mind in space and time but in space and time in the mind—how they have arisen there and how they have changed and developed in the course of mankind's biological and psychosocial evolution.

To inquire into the sources of our spatial and temporal notions we must turn to their origin and development in anthropogeny and, from the positions of developmental psychology, analyze their genesis in the process by which different types of intelligence develop and succeed one another in the child. Pathopsychological materials are of some assistance here, for the decline of higher mental faculties is accompanied by a regression in spatial and temporal conceptions. Paleopsychology and an analysis of the mythological world view will help us to approach these notions as they evolved in primitive culture, ethnogeny, and religion. This investigation will facilitate an organic transition to the history of philosophy and science, where we will encounter an entire spectrum of temporal concepts, many of which are still relevant today.

Before proceeding to this study, however, I would like to observe that its value is not just historical or psychological; it also contributes to the development of materialist epistemology. Lenin stressed that the theory of knowledge and dialectics must encompass many areas: the history of philosophy, the history of individual branches of science, the history of the intellectual development of the child and of animals, the history of language, psychology, and the physiology of the sense organs.[3] To a greater or lesser degree, all these areas fall within the scope of the present chapter.

Human Evolution and the Origin of Spatial and Temporal Notions

My first task is to determine how *Homo sapiens* began the epistemological structuring of nature and to trace the genesis of the spatial and temporal view of the world. A wide variety of structures has been proposed on the basis of this productive approach; many of these have long since been forgotten, but it may be that their true value will not be appreciated until we possess an adequate scientific theory. The aim here is not only to ascertain the specific features of these "seminal" structures but also to analyze possible methods of reconstructing the spatial and temporal conceptions that have arisen and figured in human evolution.

Certain scholars proceed on the assumption that the notion of three-dimensional space is inherent in humans (as in all animals), "in our blood," a preprogram underlying the subsequent structures that arise at various stages of ontogeny. In this view, substances and living beings have been deeply marked by several billion years of evolution in a three-dimensional world, and the survivors in this process have been those that have best adapted to three-dimensional space.[4] Here may be mentioned numerous ideas positing the existence of certain innate regulatory operators, perceptual structures, instinctive notions, archetypes, and so on.

Naturalists and philosophers of different persuasions have devoted increasing attention to such primary structures. Their ideas about how these structures can be known differ accordingly. Thus existentialists, for example, Heidegger and Merleau-Ponty, and phenomenologists, such as Husserl, Gurvich, and Schutz, consider that, because aesthetic, religious, and scientific relations and structures have been erected on mankind's original relationship to the world, we will not be able to penetrate the secret of that relationship until we have rid ourselves of the modern forms of rational scientific thought.[5] Through transcendental reduction (which is not unlike religious revelation), phenomenology hopes to arrive at the pre-logic of the perceptual world by which the totality of all primary structures can be known.[6] I find such an approach unsuitable; my inquiry into sources is undertaken in the interests of science and with the help of science, whereas that of the phenomenologists regards science and its interpretations as a distortive factor that must be eliminated. What we need is not a "complete" knowledge of primary structures obtained through extra-scientific methods but a partial knowledge acquired by means of scientific reconstructions whose synthesis can enable us to recreate the theoretical wholeness of these structures in accordance with dialectical materialist methodology. We are striving for a scientific picture of the prescientific universe, which includes the spatial and temporal conceptions that came into being in that world.

P. N. Anokhin has made some interesting observations relevant to the present context. Proceeding on the general philosophical thesis that the spatial and temporal continuum of the motion of matter is an absolute law that was in effect long before the appearance of life on earth, Anokhin concludes that only adaptation to the framework prescribed by it can guarantee the survival of living beings. Here we can appreciate the full fruitfulness of a well-known position of dialectical materialism as formulated by Lenin:

If the sensations of time and space can give man a biologically purposive orientation, this can only be so on the condition that these sensations reflect an *objective reality* outside man: man could never have adapted himself biologically to the environment if his sensations had not given him an *objectively correct* presentation of that environment.[7]

Anokhin observes accordingly that this reflection of the spatiotemporal continuum in living beings became an absolutely indispensible prerequisite for prediction. It deserves to be noted further that prediction is an important ingredient in the development of temporal conceptions because it is connected with the conditions for reactions that are needed only in the future.

Anokhin regards prediction as a universal phenomenon of an organic nature that is based on spatiotemporal rhythmicity.

Evolution, beginning with primitive protoplasmic "prediction," perfected the process to such a degree in cerebral matter that the brain developed into an organ capable at any given moment of its activity of combining the past, the present and the future.[8]

As living beings became psychically more advanced through a differentiation of the nervous system and special organs of sense and perception, there was a concomitant evolution in the reflection of the spatiotemporal continuum. All this helps us to understand the "animal" roots of anthropogeny.[9] We need not descend many rungs on the evolutionary ladder to note, for example, that the eyes of the lower monkeys, as compared with those of the lemurs, shift from the sides to the front of the head, thereby facilitating binocular vision of the surrounding world. Now binocular vision is not only organically connected with adequate spatial perception but also required for temporal perception. Indeed, V. Favorskiy concludes in his analysis of aesthetic perception that "binocularity contains time in a very condensed form."[10]

The development of bilateral mechanisms (binocular vision, binaural hearing, bimanual touch, etc.) for constructing the images of external objects is an important topic deserving special attention, as this evolution is contingent on basic properties of surrounding space, including three-dimensionality. We should also note the bilateralization of conditioned (as opposed to unconditioned) reflex mechanisms. According to B. G. Anan'yev and Ye. F. Rybalko, this is a phenomenon of fundamental significance, because it is connected with the orientation of the organism in the space of the environment. They advance the interesting hypothesis that the special adaptation that higher organisms have achieved to the spatial conditions of existence is accounted for by the paired activity of the cerebral hemispheres.[11]

The appreciable development of discriminatory functions that occurred in the auditory apparatus of the monkey was an important prerequisite for signaling in flock life and for the coordination of group spatial movement and orientation. Kinesthesia, especially of the hand, was significant in the sensory evolution of the monkeys, because this and vision are the basis of perceptual organization and vital ingredients of spatial perception.[12]

Thus motility and vision are the two means by which humans have mastered the two basic (topological and metric) properties of space. Observing in his studies on biomechanics that the motoric field exhibits topological categories, N. A. Bernshteyn advances the following interesting hypothesis:

> Locally reflected in the higher motor center of the brain we find nothing but a kind of projection of external space itself in the form in which it is given to the subject [not musculoarticular schemata—M. A.]. From the foregoing it follows that although this projection must be congruent with external space, it is a topological rather than a metric congruence.[13]

As to the metrics of space, they are closely connected with vision and the refractory apparatus of the eye.

The appearance of labor activity, language, and thought is organically related to the coordination of vision and the developed kinesthesia of especially the hands. The brain, especially such areas as the frontal, lower parietal, and parietotemporo-occipital subregions involved in such specifically human activities as language and graphic skills, evolved rapidly in anthropogeny. In contrast to the higher nervous activity of the anthropoids, which is biological in nature, human labor and thought are social phenomena[14] whose qualitatively different relationship to the surrounding world presupposes a different adaptation to space and time. Through labor mankind exerts an active influence on nature and begins to comprehend the essence and structure of objects and processes in objective reality. On this boundary are laid the first building blocks of the majestic future edifices of arithmetic, geometry, and logic. The logical character of developing thought is based on nature and labor. The differentiation of spatial relations provides for the functioning of a second signaling system. As V. I. Kochetkova notes: "The parallelism that often can be observed between the development of techniques and the presence of certain brain structures in human fossils indicates that these phenomena are interdependent."[15] These parallels may help to explain the evolution of spatial and temporal conceptions in anthropogeny. Here, received data must be correlated with the structure of the primitive social aggregate, the distinctive features of paleolithic cultures, the structure of archaic practices, the development of primitive thought and language, and mystical and symbolic notions that arise at a certain stage of perception.

Another factor to be taken into account is the extremely significant advance in the way time is reflected in evolving living matter. For example, although the "biological clock" also operates on the level of unicellular organisms, such mechanisms as the perception of temporal sequence and distant temporal perspectives, the measurement of time, and the representation of the abstract notion of homogeneous and continuous time are unique human characteristics that have resulted from a long process of cognitive evolution.

Schools such as Kantianism and nativism have long regarded space and time as innate forms of perception. In the nineteenth century the source of our spatial sensations was an important philosophical and epistemological problem central to all sciences about the external world. The physiology of vision as developed by such scientists as Müller and Helmholtz brought about a radical change in approaches to space and time. Attention began to shift from general questions on the origin of these forms of perception to certain specific physiological and psychological problems. The study of these demonstrated, for example, that "the capacity of the adult retina for spatial vision is not innate but acquired through experience."[16]

Thus, even if certain geometrisms and rhythms are "in our blood," it is only with "our mother's milk" in early childhood that we acquire our first information about space and time. Naturally, due attention must be given to both ontogenetic and phylogenetic childhood. Such an approach is all the more justified in view of the fact that the child's intellectual development reproduces in a specific and condensed form the intellectual evolution of mankind in general.[17]

The sense organs of the newborn infant are fully operational, but initially only the most elementary (cutaneous and gustatory) senses actually function. They enable the infant to take its very first bearings in space by finding its mother's breast, etc. As to vision, the newborn does not really seem to see at first. It is unable, for example, to direct and focus the uncoordinated movement of the eyes, which even move independently of each other. The child still has to learn to see. It is important to emphasize here that eye movement is regulated in response to moving external objects. Thus, by the second to fourth month, the infant's eyes are sufficiently developed to allow it to follow an object. As D. B. El'konin observes, at this age eye movement is evoked by the movement of an object.[18] The eyes themselves do not yet examine or seek out the object; this comes later in connection with the development of kinesthesia of the hands and speech comprehension. The child has not yet reached the stage of sensorimotor intelligence, but its eyes grasp movement in accordance with a group of movements of macro-objects in three-dimensional space. B. G. Anan'yev and Ye. F. Rybalko have demonstrated that a moving object is the first and most basic prerequisite for the development of spatial perception.[19] They consider that it is all the more important to direct attention to this observation because it refers to the period when the movement of the child itself (even grasping) cannot yet play any role whatever in the reflection of external influences. True, even during this initial period the child is not immobile but is rather a dynamic field of reflexive and physiological movements (searching for the breast, sucking, crying, peristalsis, beating of the heart, etc.), which are of course influenced by the external world and are connected with Ashby's three-dimensional preprogramming geometrism. Primary space is this system of reflexive and instinctive relations.[20] The system is biologically universal because it is determined by such fundamental physical and cosmic factors as the gravitational field of the earth, solar radiation, and heredity. This prespace serves as the foundation for subsequently erected structures, such as visual mechanisms and the nervous system, that are specific to different forms of life and dependent on the distinctive features of their ontogeny and ecological environment.

The surrounding environment is not universal; instead, different species live in different worlds. It is particularly important to take cognizance of this fact in comparative investigations. In studying how perfectly various living species have adapted to their surroundings, for example, if we postulate the same environment for the investigator and all other organisms that have achieved some measure of adaptation, then our human environment becomes the only standard; the structure of the lower animals appears inferior to that of the higher ones. Biologists such as von Uexküll have objected to this approach. Relative to its environment, no animal is "more or less" adapted—all have adapted equally perfectly. A similar state of affairs arises in the analysis of genetic problems in psychology. Thus Albert Peiper correctly observes that serious errors will inevitably occur if, as often happens, the subjective environment of the adult is presumed to hold for the infant.[21] On the other hand, it must be remembered that all our interpretations of the child's world are necessarily those of an adult.[22] This geneticopsychological analog of Bohr's epistemological principle will be kept in mind in my discussion.

At the initial formative stages of infantile thought—reflexes and elementary habits—there is no perception of a single space. The individual exists, as it were, in diverse and heterogeneous spaces corresponding in number to the different perceptual fields (gustatory, visual, etc.). These separate spaces, however, also undergo changes, exhibiting a gradual tendency to unite or synthesize, because the perception of a unified space is a complex intermodal association that is systemic in nature.

Let us consider the development of spatial perception. First, there is a gradual increase in the dimensions of visually perceived space. Initially the child encounters an image of undifferentiated, amorphous continuity; it then begins to distinguish discontinuously (irregularly or jerkily) moving objects. This initial perception merely reflects the boundaries or breaks between a solid spatial mass and an individual object distinguished from it by movement;[23] that is, what we are actually dealing with is a unidimensional image with the contours of a moving object. This is followed by a stage at which the perception of external objects is more comprehensive but still lacks perspective and is on one plane. This is the level of two-dimensional space. Finally, three-dimensional perception appears at the level of sensorimotor organization.[24] Here we must observe that, although the child has attained a three-dimensional perception of the surrounding world, it is not yet able to render depth when it attempts to portray what it sees. Interestingly enough, painting in certain ancient cultures, such as the Egyptian, exhibits similar features.[25]

This stage is followed by a gradual expansion in which the primary space whose image is formed with the help of the mouth (thus I am using more than a simple figure of speech when I say that the infant receives its first information about space with its mother's milk) is extended to larger spaces that take shape in connection with the dynamics of the hands, accommodation of the eyes, etc.[26] The child broadens its visual space appreciably during the first months of life. At the age of four to five weeks it can follow a moving object at a distance of 1–1.5 meters; at the age of two months this has increased to 2–4 meters; and by three months the distance is as much as

3–7 meters. As its auditory orientation comes into play, the child also begins to focus on unseen space.

As to the conception of a unified space (even a primitive one), this does not appear until the development of sensorimotor intelligence. It is on this level that objects are perceived as something invariant, and elementary spatial systems begin to form as vision and prehension are coordinated. Spatial representations develop on the basis of a progressive organization of movements, which almost form a group structure. Within the limits of this intelligence, however, it is impossible to attain more than the representation of near space; the sensorimotor group is only a schema of behavior that never acquires the status of an instrument of thought.[27] At subsequent stages of intelligence, thought reaches the level of conceptual intelligence. If sensorimotor intelligence operates only with concrete materials in near proximity to the subject, conceptual intelligence aspires to embrace the totality of the surrounding world, and the spatial distances between subject and object increase to infinity.

Perceptual space is heterogeneous and possesses a certain centration. The development of the child's mental processes is accompanied by a transition from a general egocentrism to an intellectual decentration. Such constructions as space, time, causal connections, and the empirical group of object displacements are created as the child liberates itself from its perceptual and motoric egocentrism. These successive thought decentrations are followed by the attainment of the level of operational groupings and thought groups. The difference between sensorimotor and conceptual intelligence here not only consists of a quantitative increase in the spatial dimensions of the surrounding world and greater distances between subject and object but also includes a qualitiative distinction in the structure of space itself.

It must, of course, be borne in mind that, between these two levels, which in fact characterize both the lowest and the highest stages of intelligence (the group of sensorimotor displacements and the operational and formal groups, respectively), there are several intermediate stages, such as preconceptual thought and intuitive (representative) thought. The transition to each successively higher level is accompanied by a change in the structure of the space within whose limits the subject perceives the world. The mastery of language achieved at the preconceptual level, for example, depends on the child's ability to use signs as symbols and to represent one thing by means of another (such features are totally alien to sensorimotor intelligence). Preconceptual intelligence is characterized by preconcepts, or participations. The primitive inferences uniting such preconcepts follow from immediate analogies, rather than from deductions, and are known as transductions. On this level the child is still unable to distinguish general classes, because it lacks the distinction between "all" and "some." On the other hand, the creation of a representation of an object in the immediate field of action does not mean that an analogous concept arises that would embrace a large space or repeated appearances of the object at given time intervals.[28]

Such features of preconceptual intelligence indicate that a unified, continuous space in the small becomes discontinuous and multiply connected in the large, and participations arise among distant objects that strongly resemble the characteristics of primitive thought as described by Lévy-Bruhl. The child's world on the level of preconceptual intelligence can be said to be mythological. It is significant that we entertain children of this age not with deductive, scientific constructs but with stories set in a make-believe world ruled by mythological space and time. After all, the child does not yet have a unified space and time; what it does have is a stratified space of different worlds linked extraspatially and extratemporally by wells, tree hollows, and the like. The drawings of small children are also interesting in connection with preconceptual thought. Consisting of isolated parts, such pictures indicate that the child lacks the concept of a single space. In the pictures we can trace some of the stages in the development of a unified space concept and follow how it gradually assumes sharper and fuller contours.[29]

What of the development of temporal conceptions? From the moment of birth the infant is influenced by sensations of certain temporal relations, circadian rhythms, etc. As to its conditioned reflexes to time in their various aspects, these are forms of adaptation common to all animals.

The idea of the future is born in the first days of life. As J. M. Guyau puts it: "When the infant is hungry, it cries and stretches out its hands to its nurse: here you have the idea of the future in embryo."[30] One stage in the mastery of time is connected with auditory orientation, because, as I. M. Sechenov emphasizes, "hearing is the analyzer of time."[31] Although there are a great many such temporal aspects in the life of the newborn, the child is late in acquiring a concept of time. This may be due to features of the evolutionary process, in which a sense of space precedes one of time. It can also be observed that, as the child attempts to conceptualize time, he or she actually operates with spatial relations for a considerable period.[32] The same features have set their stamp on the history of language: Words designating spatial relations are used to express temporal relations, temporal prepositions are developed from spatial ones, and so on. We must remember, of course, that in many developed languages the difference between past, present, and future is clearly reflected in the verb; in other words, the idea of time is in a sense imposed on us by language itself. The significance of this as a means of mastering time, however, should not be exaggerated. In the first place, the child's earliest speech lacks inflections and auxiliaries. As Peter Lindsay and Donald Norman observe: "The child seems to talk only of the present: about the actions that are going on around him and the objects he is seeing."[33] Second, although the child has a mastery of language on the level of preconceptual thought and although temporal words, such as "today," "tomorrow," and "yesterday," have appeared in its speech by the age of three years, such words are for the most part used indiscriminately.[34]

The preceding, I think, seems to oblige us to consider the Sapir-Whorf hypothesis of linguistic relativity, not, however, as it is traditionally employed to discuss the comparability of different languages and the worlds they express but rather with reference to speakers of the same language on different levels of intelligence. Here a question arises that seems superficially paradoxical but is actually quite profound: Do we and our children speak the same language? Because physics utilizes a great many concepts whose meaning has changed significantly as different theories have succeeded one another, a similar question presented itself there when classical theory was compared to the later constructions of relativistic and quantum physics. This has led certain observers to doubt the commensurability of scientific paradigms and to question whether scientific theories are really cumulative. This central problem in contemporary logicomethodological studies of the history of science and the dynamics of physical theories derives from essentially the same phenomenon we encounter in communicating with our children. The world of the adult and the world of the child are distinct and operate with different languages. They both use the same words, but the level of concreteness or generalization, the symbolism, and the space, time, and causality expressed by them differ considerably.[35]

As to the sequence in which space and time are mastered in language, it apparently can be found in a general form in the transition from rhetoric to grammar described by Michel Foucault:

Rhetoric defines the spatiality of representation as it comes into being with language; grammar defines in the case of each individual language the order that distributes that spatiality in time. This is why . . . grammar presupposes languages, even the most spontaneous and primitive ones, to be rhetorical in nature.[36]

I noted that in preconceptual thought universal space appears as a number of ambiguously connected, locally continuous spaces. Multiple local times are also characteristic of this level: Each change has its own time, and their interrelationship is as yet discerned vaguely. Intuitive time, as analyzed by Piaget, displays similar features.[37] In intuitive (representative) thinking, time is connected with individual objects and movements; it is not homogeneous, nor does it flow uniformly. Thus, if two bodies simultaneously begin to move in different directions and at different speeds away from a point A and then stop simultaneously at points B_1 and B_2, respectively, a four- or five-year-old child will be unable to grasp the simultaneity of their arrival, even if it is easily perceptible. The child refuses to understand that the movement ended "at the same time," because he or she has no concept of a universal time for different speeds (and in this respect the child is closer to Einstein than to Newton). "Before" and "after" are distinguished as a spatial rather than a temporal sequence. To the child, "faster" means "farther," and this involves "more time." Even when temporal measurement becomes possible at the "operational stage," a seven- or eight-year-old child will still be unable to perceive time as a relation independent of concrete changes. The child is willing to believe, for example, that the course of time can be influenced by moving the hands of a clock. As P. Fress observes, an abstract concept of homogeneous and continuous time signifies a high degree of adaptation to change and does not occur until the threshold of adolescence, when formal operations develop in response to the demands of experience that time be distinguished from changes.[38]

To conclude my analysis of the stage-by-stage development of the child's perception and representation of space and time, I should note that the reverse process can also be traced in the disintegration of spatial and temporal structures of perception encountered in various psychopathologies. This is an important area of research. First, the psychopathological destruction of perceptual spaces and times is by no means merely the reverse image of the ontogenetic acquisition of spatial and temporal perception. A number of other studies have noted that analyses of psychic disturbances can contribute a great deal to an understanding of normal psychic activity. I will analyze the features of the perception of time and space (and features of perceptual times and spaces) that have remained beyond the scope of geneticopsychological investigations.[39] Second, some researchers[40] regard psychopathology as something of a throwback; that is, they consider psychic anomalies, which are encountered relatively rarely today, to be phylogenetically atavistic.

Many pathologies indicate that the disintegration of the psyche is a step-by-step process. The first strata and formations to suffer are the youngest and genetically most recent, which release certain normally "submerged" mechanisms. On the other hand, James and others have proposed that the paired activity of the human cerebral hemispheres is a relatively recent development of the last few thousand years or so and that earlier human species had a "bicameral" mind. This suggestion is of interest to both pathopsychology and the study of archaic conceptions and mythology. Third, an analysis of spatial and temporal perception in psychopathology is important in making the transition to a study of primitive and mythological conceptions because such notions were enormously influenced by the shaman or wizard, usually a psychically abnormal individual whose mystical and magical actions always repeated the first attack during which his calling was revealed to him.[41] Specific psychosomatic states were generally typical of individuals in primitive societies (rites of initiation, dreams vaguely distinguished from real life, etc.).[42] All this helped to create an intellectual atmosphere conducive to a world image in which a prominent part was played by psychopathological features of spatial and temporal perception (stratified spaces, bipresence, extratemporal wanderings about the world, etc.).

Modern pathopsychology has amassed an enormous body of material[43] dealing with the changes that occur in spatial and temporal perception and representation in various psychic anomalies caused by damage to specifically human parts of the brain (for example, disturbances in the commissure, such as commissurotomia). Thus lesions to the frontal lobes disrupt the unified spatiotemporal organization of perception; injury to the temporoparieto-occipital subregion is accompanied by disturbances in the complex forms of spatial analysis and synthesis; the properties of time are greatly distorted by diffuse organic lesions to the cortex—the patient feels that time has stopped, is unable to relate "his" time to time outside him, time loses its direction and becomes convoluted; damage to the temporal regions erases the distinction among past, present, and future. Changes in spatiotemporal representations, disturbances in chronogeometric orientation, and stratification of the space and time of perception characterize such psychic illnesses as the Kandinsky syndrome, senile psychoses, schizophrenia, and epilepsy.

This extensive body of data has yet to be classified. As T. A. Dobrokhotova and N. N. Bragina note: "The extant specialized literature contains practically no systematized presentation of disturbances in the perception of time and space, although there are many scattered references to such disorders."[44] In the future we will probably have to resort to a logicomathematical and physical interpretation of this material. A great deal of work will be needed to determine the specific metric, topological, group, and other structures of the spaces and times within which the worlds of different psychic pathologies operate. However, the study of pathological states can already teach us a great deal about how the space and time of our perceptions and representations are structured.

The clinical picture of focal lesions to the brain, for example, led to the notion of functional asymmetry. This was followed by the question of the relationship of the left and right hemispheres to time, as this is essential to determining the basis of sensory and abstract cognition. Left/right agnosias are interesting in this connection because they indicate that bilateral vision, or the paired functioning of the visual analyzers, is by no means a simple binocular superimposition of the monocular fields of the left and right eyes. When one eye is closed, visual space is limited to the monocular field of the open eye: Spatial scope is restricted, and depth is not focused as sharply, but there is no change in structure. The spatial symmetry of right and left within the monocular field, for example, is unaffected, even though the fields themselves are unequal, one of them being connected to the dominant cerebral hemisphere. The situation is quite different when one of the functionally asymmetric hemispheres is damaged, affecting the performance of the corresponding (contralateral) eye. Thus a lesion to the right hemisphere is often accompanied by a left/right agnosia that disturbs the spatial symmetry of the perceived world by canceling sinistral space.[45] An individual with this affliction visually and motorically ignores half of space not only as a whole but also locally within the remaining half. For example, the afflicted individual will draw a daisy with petals on only one side, read only one half of the text on a page, and wash only one side of the face.[46]

Thus left/right agnosia indicates that the spatial organization of visual perception is complex. The normal combined activity of the hemispheres obscures this specific organization with respect to left and right. In pathologies, such asymmetry is vividly manifested and reveals the natural predominant tendency of right-handed persons with focal lesions to ignore sinistral space, prompting researchers to propose that the right hemisphere has a special function in visual-spatial perception. Thus V. I. Korchazhinskaya and L. T. Popova conclude in their study of spatial agnosias:

Peculiarities in the way in which defective spatial perception is compensated by verbalization suggest that besides verbal thinking, which is dominated by the left hemisphere, there may exist a non-verbal, representative visual-spatial thinking in which the right hemisphere seems to dominate.[47]

Dobrokhotova and Bragina, who have especially emphasized the functional asymmetry of the brain in their analysis of spatiotemporal factors in human neuropsychic activity, devote a fruitful discussion to this question.[48]

The essential point here is that the paired activity of the functionally asymmetric hemispheres is characteristic of the highest form of spatial and temporal reflection. The very division of the anterior brain into two hemispheres is a fairly recent development in the evolution of living matter and is typical only of the terrestrial vertebrates. Further evolution was accompanied by a considerable increase in both the size of the hemispheres and the power of connecting commissural systems, such as the corpus callosum, which attains its maximum in humans.[49] Also, the human brain is functionally asymmetric, which in the opinion of most observers distinguishes it from that of all other representatives of the animal kingdom (although here as well there is a certain asymmetry in walking, running, flying, etc.). The functional structure of the brain has a complicated dynamics in human ontogeny. The brain of the newborn is for a time functionally symmetric (this symmetry may persist throughout the individual's life, being typical of certain forms of dementia). Functional asymmetry is not noticeable until the second year of life, after which it continues to increase, attaining maximum development in adulthood parallel to a similar progression in the paired activity of the hemispheres but declining in old age. This asymmetry merits close attention. Here we are concerned with far more than simply determining the mechanism by which perceptual space is organized with respect to left and right, although this in itself is an interesting problem. The incomparably greater question, however, has to do with the sources of the divorce between the theoretical and the empirical in human psychic organization, as each hemisphere is responsible for different forms of thought. Thus imaginal, visual-spatial thought is controlled by the right hemisphere, whereas we find both abstract, logicoverbal thought and most specifically human functions (such as linguistic and mathematical competence, speech, and writing) controlled by the left hemisphere. Dobrokhotova and Bragina's analysis of focal lesions to the hemispheres led them to study the intriguing problem of the part played by real space and time in sensory and abstract cognition processes. I agree wholeheartedly with their observation that "it is surprising how late this question has presented itself, how long it has been ignored by both the extant theories of cognition and the psychopathological literature."[50]

What is the role of space and time in sensory and abstract cognition? Sensory cognition processes are organized in concrete space and time because the formation of sense images necessarily demands immediate contact between the subject and the object of perception through visual, auditory, tactile, and other stimuli, and this always occurs in concrete space and time. Such organization is evidently taken care of by special mechanisms in the right hemisphere, which is also responsible for the formation of the sensation or experience of space and time as a relatively independent form of perception or sensory cognition. However, there is also another, uniquely human kind of cognition that consists in the ability of humans to exist in real space and time and yet liberate themselves from their confining limits. This is achieved through abstract cognition, which operates only with symbols—words, speech, signs, and the like—of real objects and phenomena.

In all probability, abstract cognition could not develop until a special mechanism had appeared in the brain which could both keep the continuous psychomotoric activity of the waking individual in contact with real space and time and yet, by freeing the process of verbal (abstract) cognition from that same spatial and temporal framework, allow him to step outside its bounds. Such a mechanism is evidently located in the left hemisphere.[51]

The left hemisphere is associated not with concrete space and time but with the abstract space and time of concepts. With respect to time, we cannot rule out the possibility that the respective hemispheres are, in general, oriented toward different times: The right hemisphere contains all the experience of sensory cognition accumulated in the past and in its functioning is connected with the present and the past, whereas the left hemisphere, associated with abstract cognition, generates programs for the future and is thus linked to the present and the future.[52]

The division between abstract and concrete spaces and times and the difference in the temporal orientations of sensory and abstract cognition are best observed in various forms of cerebral pathologies, because in the normal paired functioning of the (adult) brain the two hemispheres operate simultaneously to provide unified imaginal-conceptual models of the real world. These integrated empiricotheoretical sensory/abstract constructs are presently being given considerable attention by psychologists and philosophers. The many factors that account for this interest include the increasing fertility of dialectical methodology in modern science and the bankruptcy of the positivist program that clearly divorces scientific theory and observation. Thus psychology is considering an imaginal-conceptual model deriving from complex creative activity on the part of the subject. V. P. Zinchenko, V. M. Munipov, and V. M. Gordon, for example, come to the interesting conclusion that this model is a multidimensional reflection of reality described in various perceptual, symbolic, and verbal languages.[53] In my opinion, however, what we are dealing with here is instead a multilayered reflection in a stratified space: The superimposition or "lamination" of the strata (each of which, generally, must have its own specific structures of relations) is what provides us with an integral picture of reflected reality. Such structures are determined by the objective makeup of that reality and specific features related to society and the internal organization of intelligence. Problems meriting attention here include determining the number of such strata, their specific structures and languages, and how they are "stuck together." Historically and in psychic anomalies, all these characteristics undergo significant changes in both ontogeny and phylogeny, which brings up the important question of the interrelationship between the structural set of the subject and the structure of relations of the surrounding world.

It is not enough merely to state that normal spatiotemporal conceptions or sensations are disturbed in the mentally ill. We must also analyze the structure of the space and time in which the patients find themselves. Furthermore, we must attempt to determine how the individuals suffering from various psychic anomalies pass from our ordinary perceptual times and spaces possessing their own metric, topological, and other features to other times and spaces whose characteristics are often unusual. If we can solve this problem, the structure of the space and time of perception may conceivably serve as a criterion for the classification of various pathopsychologies. As Anan'yev and Rybalko observe, at any rate, "in future scientific diagnostics the level and peculiarities of spatial perception may well become an indicator of infantile and adolescent intellectual development."[54]

The problems discussed are not just relevant to one important current in pathopsychology; they may also prove valuable to epistemological and logicohistorical studies of such subjects as the division and unity of sensory and abstract cognition on the level of human psychic organization and the origin of spatial and temporal conceptions. For instance, if we allow for some correspondence[55] in the structural aspects of human phylogeny and ontogeny, as noted by Holl, Werner, Simon, and others, on the basis of the materials at our disposal, we can at least tentatively attempt to compare the spatial and temporal structures described by developmental psychology and pathopsychology with certain stages in anthropogeny and ethnogeny. This may contribute a great deal to the study of ancient mythology, mythological space and time, primitive thought, mystical notions, the origin of religious concepts, and archaic cosmologies.

The Evolution of Spatial and Temporal Conceptions in the Early Stages of Human Culture

I would not venture to suggest that the spatiotemporal conceptions that form and function at different levels of the developing (or deteriorating) intelligence of an individual from an established social group coincide with those concepts that have taken shape in human phylogeny and ethnogeny. An analysis of the sources of such notions from the positions of developmental psychology and pathopsychology, however, provides us with the foundation necessary to examine more complex concepts of a sociocultural origin treated in historical psychology (L. S. Vygotskiy, E. Meyerson), metamythology (L. Lévy-Bruhl, C. Lévi-Strauss), and other disciplines. Here, of course, we must from the outset address ourselves to the question of whether primitive thought differs from that of the modern European, and if so, whether the distinction is qualitative or quantitative, on the level of thought processes or thought content. Thus, on the one hand, Franz Boas considers that "the functions of the human mind are common to the whole of mankind,"[56] and Lévi-Strauss[57] firmly declares that the principles on which reason functions are the same in all cultures and all ages. On the other hand, a majority of observers, including Lévy-Bruhl and Vygotskiy, conclude that mental processes exhibit significant intercultural differences. It is important to understand the issue of this dispute, because if the advocates of the first viewpoint are right, little is to be gained from comparing the spatiotemporal conceptions that arise at different stages of phylogeny and ontogeny (whose levels of intelligence have been precisely described by developmental psychology). If intelligence is universal, then we must conclude that spatial and temporal conceptions are also universal and subject only to insignificant and purely external changes from one culture and period to another. Without denying the presence of certain universals that structuralists tend to absolutize, I adopt the second point of view. I choose to do so not only because there is a large body of material that speaks in favor of this viewpoint (it would be ridiculous to think that a scholar of Lévi-Strauss's stature would base his conclusions on a small or unreliable collection of data[58]) but also because it is intrinsically consistent with basic tenets of Marxist-Leninist philosophy. This last point deserves special emphasis, as the answers to the questions confronting us here can be provided only by philosophy and natural science working in close collaboration. Leroi-Gourhan may consider that the only real significance of "studies of prehistory—whether they are propped up by metaphysics or dialectical materialism—is to situate future man in his present and his remote past,"[59] but the philosophical "props" chosen can produce radically different models of the psychological aspect of human phylogeny. In the one case we presume that intelligence is universal in all ages and cultures (a divine gift perhaps?), whereas in the other it is regarded as arising in a changing sociocultural environment!

According to the notion developed by Friedrich Engels,[60] human consciousness is immediately bound up with the appearance of labor (and with it, articulate speech), on the basis of which human society develops. Characteristic of labor is the mediatory role it plays in the development of society and the means of production. In *A Contribution to the Critique of Political Economy,* Marx wrote:

In the social production of their existence, men inevitably enter into definite relations, which are independent of their will, namely relations of production appropriate to a given stage in the development of their material forces of production. The totality of these relations of production constitutes the economic structure of society, the real foundation on which arises a legal and political superstructure and to which correspond definite forms of social consciousness. The mode of production of material life conditions the general process of social, political and intellectual life.[61]

The development of human consciousness is vitally dependent on social being. But because sociohistorical progress occurs in qualitative leaps rather than as a quantitative evolution, the development of the human psyche and consciousness may at certain stages also take the form of qualitative change.[62] A majority of scholars guided by the methodology of dialectical materialism (L. S. Vygotskiy, A. P. Luria, A. N. Leont'yev) or by ideas close to Marxism (H. Wallon, E. Meyerson) take just such an approach, which does not, of course, exclude other paths leading to the notion that thought differs qualitatively at different stages of cultural and social development.

Thus human thought undergoes qualitative changes in the process of sociocultural evolution—hardly a surprising fact, considering what has been presented in the preceding section. Here, however, the historian and the psychologist are confronted with the following important problem: How are we to penetrate the thought of distant ages, how can we know the conceptions and values that were dominant then? In the most general terms the answer can be stated in the spirit of historical psychology: The history of human thought can be known through the history of mankind's creations (labor). As to specific methodology, I. Ya. Gurevich, for example, formulates it when he says:

We assume that the proper approach is to attempt to discover the basic universal categories without which culture is impossible and with which all its creations are permeated. These are at the same time the determinative categories of human consciousness. Here are meant such notions and forms of perception as time, space, change, cause, fate, number, the relationship of sensory to supersensory and of part to whole.[63]

Each culture constructs its model of the world from these cosmic categories, which are contained in sign systems, such as language, mythology, religion, and science. Gurevich stresses that he regards these fundamental conceptual and sensory categories as universal because, although their content varies, they are inherent in humans at any stage of their history. This is why studying these categories is of such paramount importance to an understanding of the culture and social life of different historical periods—which brings us to the next question: Can modern humans adequately comprehend archaic categories? After all, there is a real and perhaps unavoidable danger that we may fill them with a new and alien meaning. These considerations lead Gurevich to a more general problem: "Are we capable of understanding the past without imposing upon it the view dictated to us by our own environment and age? It would seem that this question must be answered in the negative".[64]

An integral approach appears to offer a key to this problem. In one of my earlier works[65] I outlined three areas of study—developmental psychology, pathopsychology, and primitive thought—that might help to explain the sources and evolution of spatial and temporal conceptions. The subject of study in all three cases is the structure of other worlds. Modern science is familiar with this state of affairs. Thus in physics macroscopic observers study the microworld on the basis of macroscopic data and paint their picture of the micro-object with a "macro-brush" of an appropriate physical theory. This is basically what Niels Bohr meant by his famous epistemological principle: "Regardless of how distant phenomena are from the framework of classical physical explanation, all experimental data must be described in terms of classical concepts."[66] We encounter a similar situation in the study of the infantile, psychopathological, and archaic worlds under consideration here. We can offer only an adult interpretation of the child's world (let us remember the wise aphorism that understanding the atom is child's play compared to understanding the child's play), a normal interpretation of the pathological world (for no matter how insistently Karl Jaspers may urge us, we will never be able to observe, inquire, analyze, or think psychopathologically), and a civilized (Eurocentric, scientistic, etc.) interpretation of primitive thought and ancient mythology. Whenever we attempt to transfer infantile, psychopathological, and archaic conceptions onto our modern notional grid, the worlds being investigated and the world of adult, normal, and civilized interpretations force us to omit or distort a great deal. Taken together, however, the three complementary interpretations can aid us in constructing many of the spatiotemporal conceptions that have developed in the course of human genesis.

Let me turn first to the mythological world image and attempt to determine its characteristic spatiotemporal conceptions. I will not be concerned with the essence of myth itself, because Marxist philosophy and specialists in folklore and literature have already subjected the enormous variety of theories and studies on the topic to a thorough critical and comparative analysis.[67] The main point for my purpose is the following: "With regard to myth only one fact is indisputable: myth is narrative which, however improbable it may seem to us, was taken for the truth where it arose and gained currency."[68] Today we discover in myth an entire spectrum of ingredients corresponding to the classifications prevalent in our culture; myth is allegory, poetry, symbol, science, philosophy, archetype, structure, and world view. Myth is a peculiar primitive syncretism of these components, and yet it is something more, for "to its creators it was objective reality."[69]

The dynamics of the mythological world can be described in the most general terms as a transition from Chaos to Cosmos. Chaos is the initial category of mythology and the initial state of the mythological universe: "Verily at the first Chaos came to be."[70] Also, if we take not the ancient Greek mythology of Homer and Hesiod, in which we find a relatively generalized notion of Chaos and beyond it the abyss, the void, the ocean, infinite space, darkness, etc., but turn instead to archaic Australian, African, Siberian, and other mythologies, we find that the image of the primary unordered world is more concrete and generally connected with water. "In the beginning there was no land, there was but one water."[71] This phrase is taken from the mythology of the Kets, but it is typical of the overwhelming majority of cosmogonic myths (the Sumerian Nammu, the ancient Egyptian Nun, the ancient Indian Asat, the Babylonian Apsu, and others).

Reminiscences of the primacy of water can also be found in later forms of social consciousness. Thus in Christianity there is the idea that the spirit of God at first moved upon the face of boundless waters, and in the early Greek natural philosophy of Thales, water is the fundamental principle of the world. It is the amorphousness of water that lends it viability as a fundamental principle (the same applies to Anaximenes' air or Anaximander's *apeiron*). Water epitomizes the indefinite beginning which, with the help of Dike, Nous, the Idea, God, etc., is structured and transformed into spatially and temporally ordered Cosmos.

Cosmos is the culmination and final category of mythology; in the oldest mythological conceptions there is no notion of the world as a definite unified whole. On the contrary, relative order is connected only with the tribal territory, which is surrounded by and permeated with a mysterious and pernicious force that can in Lévy-Bruhl's terms be called the supernatural world. The powers at work in this world are unordered and unhierarchical, a "fluid multiplicity of vaguely related elements."[72] This is the protoimage of the ancient Greek Chaos, which is dominated by sensory and emotional elements, such as darkness, the abyss, and death.

Relative order in this narrow little world of primitive mankind (an analog of the near space that initially forms in the child) was unstable and less natural than moral and social in character, because only the observation of definite rules and taboos could ensure its preservation. Here, of course, it must be remembered that in ancient times the world was not divided into nature and society. These were united, so that the developing picture of the world was sociomorphic.[73] Thus group division within the tribe determined spatial division, and the order of ritual ceremonies was connected with spatial direction.[74] To the primitive mentality, "space does not appear as a homogeneous unity, irrespective of that which occupies it, destitute of properties and alike everywhere."[75] Supernatural, mystical forces were responsible for the different properties of spatial regions (Lévy-Bruhl once noted that our contemporary notions "supernatural" and "mystical" do not render the full essence of the corresponding primitive conceptions, but, bearing this reservation in mind, I use them for want of more adequate terms). There were a great many such objects in primitive society, but we can single out the most important or universal of them. At an early stage primitive thought was connected with totemism, so that totemic centers on the territory of the tribe were the places of greatest "mystical energy," and the space of these places accordingly possessed the most positive qualities. The space surrounding the territory of the tribe was also positive, as it was protected by the spirits of ancestors buried there. In this sense, myth was contained in the landscape itself;[76] each significant detail of the surrounding natural environment was associated with some magical force or spirit of a forebear, being either its hypostasis or its representative (which "bipresence" rendered as nearly the same thing). George Thomson notes:

> As the pair of clans evolved into a tribe divided into moieties, phratries, clans, and subclans, these totemic associations expanded likewise until they formed a cosmological system embracing the whole of the known world.[77]

Thomson's theory describes well how the sociomorphic structure of the ancient cosmological system developed, but this is still far from a unified picture of the world or, rather, a picture of a unified world. As already noted, Cosmos is a product of the mature mythological consciousness. In the primitive world view, beyond the relatively ordered space of the tribal territory (rituals, taboos, sacrifices, etc. maintain order within that territory, but mankind's world did not end there) lay an external space permeated by negative properties. The protection of the tribe's ancestors did not extend to this region, which was wholly in the power of the mighty and pernicious fluid multiplicity.

As mythology (and intertribal relations) developed further, we find that the oases of ordered existence multiply and participate mystically with each other, often to the extent that they appear to be different parts of a single oasis of being. For example, if an old woman of the tribe fell ill, the explanation was that a mouse had fallen into a snare in the distant land of mice (recall that in totemism humans maintain an identity with their ancestors and the totem). This leaves the shaman or sorcerer with two alternatives: He can either cure the old woman *here*, or he can go *there* and set the mouse free (in the *kamlaniye* ritual the shaman goes into a trance, travels about different levels of the world, visits many lands, engages in fierce hand-to-hand combat and finally releases the mouse, that is, cures the woman). These two events are merely different hypostases of a single situation or event in a world characterized by dual or multiple presence and reincarnation. This stage of mythology can be compared with preconceptual thinking. In its spatio-temporal aspect this picture of the world is similar to the representation of the child on the level of preconceptual intelligence: The space of the "local" world is continuous, whereas that of the "big" world beyond is disconnected and multiply connected.

The mythological world gradually acquires a multilayer, multilevel character. The first steps in this process have been described by Ye. M. Meletinskiy as follows:

The transition from the amorphous element of water to dry land in myths is an extremely important event necessary to the transformation of Chaos into Cosmos. The next step in the same direction is the separation of the sky from the earth—an event that may essentially coincide with the first one, since the sky is initially identified with the sea.[78]

This multistratification, however, occurs first in a plane model, a two-dimensional world, as it were, and this corresponds to the stage-by-stage evolution of spatial and temporal notions in human ontogeny. Other observers have noted this feature of primitive perception. Thus, using as her example the archaic conceptions of the Selkups, Ye. D. Prokof'yeva has shown that the "horizontal, linear" model of the world order is old, is found in only a few shamanistic rituals, and precedes a description of the "vertical" model. She observes that

in these notions the upper world is portrayed as the upper reaches of a river, the lower world is its mouth, and the middle world inhabited by man is its middle course. And since these rivers flow northward, the upper world is thought to be located in the south and the lower one in the north.[79]

The upper world was therefore the world of ancestors, and the lower world was that of the dead. This multilayer model was later extended to include three dimensions and was ordered vertically as Heaven, Earth, and Tartarus. These levels were no longer united by a river but by means of a "world tree" with roots in the netherworld and crown in the heavens and with animal mediators, such as the squirrel. The further development of the model has been described by V. G. Bogoraz (Tan): "By splitting the upper and lower of the three tiers of the world, they may be transformed into five, seven, nine tiers or worlds. The middle, earthly stratum, of course, always remains whole."[80] The multiplication of world levels increases sharply in mythologies influenced by Buddhism and Lamaism. In the mythology of the Altaic and Tuvinian peoples, for example, there are ninety-nine worlds and thirty-three layers of sky, so that here we can already speak of cosmological number mysticism.

The "fluid multiplicity" also evolves, being ordered into various negative regions or levels of the world. As the oases of ordered existence multiply and expand, Chaos as the source of evil is crowded out into the periphery; such, for example, are Jotunheim and Utgarth, the land of the frost giants in the ancient Scandinavian sagas.[81] M. I. Steblin-Kamenskiy has elucidated one important feature of mythological space on the basis of Eddic mythology:

Space exists only as concrete pieces. In other words, it is discontinuous [what is actually meant here is not the discontinuity of space, which has to do with internal metrizability, but the fact that space consists of weakly connected and ordered terminal regions—M. A.]. For this reason it is impossible to draw a map of the Eddic myths. The places mentioned in them are not situated relative to either the world as a whole or even to such parts of it as the sky and the earth.[82]

As the idea of a unified world or Cosmos develops in later mythologies, this feature disappears. When that happens, the mythological picture of the world becomes such a reliable guide that Heinrich Schliemann, for example, used Homer to discover the site of Troy.[83] As regards the model in which oases of ordered existence were located in Chaos, it seems to have figured prominently in the development of early Greek natural philosophy, serving (or putting it more cautiously, being capable of serving) as the basis of such doctrines as the Milesian school, Eleaticism, and atomism. It is clear, at any rate, that Being and Chaos are the mythological protoimages of atoms and the void, especially in view of the fact that Greek mythology specifically regarded Chaos as infinite space.[84]

The evolution of spatial notions was accompanied by a corresponding development in ideas of time. Thus the following conception was characteristic of the early stages of human culture:

We know that the primitive mind does not "sense" the successive moments of time as homogeneous. Certain periods of the day or night, of the moon's phases, of the year, and so on, are able to exert a favorable or a malignant influence.[85]

Here, mystically colored, we encounter the temporal conception of preconceptual and representative intelligence in which each change is assigned its own time and the interrelation of the two is understood vaguely. As was noted, in the multilayer model of the world characteristic of mature mythology, each level has its own time distinguished by such parameters as rhythm and duration. Thus the *kamlaniye* séance of the shaman may take several hours of earthly time, but in this short period he manages to visit several levels of the world, pursue the evil spirit, do battle with it, and return victorious to earth, that is, to the territory of his tribe. Reckoned in the internal hours of magical action, a long period may have passed—more than a human's lifetime. But after the séance the shaman stands before his fellows tired but not years older. This is because it was not the shaman himself who traveled through other worlds but his imperishable supernatural double (his soul or the like). If the body of the shaman (or any human) lives in ordinary earthly time, his soul exists in a distinct sacred time and a different, supernatural world. The soul does not disappear at death but is reincarnated in a newborn baby of the same tribe. It is this notion that underlies the oscillatory, or cyclic, model of time characteristic of mythology. I return to this model shortly, but let me first consider further the question of multiple times in the multilayer world of mythology.

During the shaman's *kamlaniye*, mankind's timeless component sets off for other worlds. What happens temporally to the shaman on such a visit? Is he still subject to the time of his world, or does he break free and live according to another temporal system of reckoning? The following typical Chukot Eskimo myth provides an answer to that question. A traveler departs for distant lands on different levels of the world, wanders about them for two or three years, and returns home to find striking changes: His home has crumbled away with age, and his children have become gray-haired and old, for many decades of local time have passed. Seeing his aged son, the young wanderer falls dead and disintegrates into dust—earthly time has claimed him.[86]

The main stress in this myth would seem to be not so much on the difference between the metric temporal features of different worlds (although this aspect is also important) as on human mortality and the inability to exist in different times. It may even be more correct to say that once a human has left his own time, return will destroy him. Thus the exit from earthly time is itself tantamount to death. Significantly, the ancient Hindus believed that a man returning from a long journey was born anew. Underlying these mythological notions is the germ of a thesis that was not clearly articulated until the advent of natural philosophy: One cannot enter the same river (of time) twice. Reincarnation may seem to violate this thesis, but in fact it is quite another mechanism that has to do not with a single river of time but rather with different temporal cycles.

A variety of mythological conceptions share one striking feature, namely, the degree to which they conform to notions in the theory of relativity. Bogoraz (Tan) perceives the source of this conformity in the fact that

> for all the ingenious complexity of its mathematical figures, in the final analysis mathematical physics arrives at a simplification, returning, as it were, to the semi-instinctive attitude of primitive consciousness before it was led astray and confused by the formal scholasticism of nominalist philosophy.[87]

From this observation Bogoraz (Tan) proceeds to an interesting and promising attempt to apply the theory of relativity and its spatiotemporal conception to the study of primitive religious notions.

The only aspect of mythological time I have considered here is the one that is intrinsically related to the distinctive features of preconceptual, representative intelligence. This aspect was not the exclusive achievement of the mythological world image—mythology merely arrays it in vivid mystical garb. But the temporal model being discussed here also figures in modern physics, particularly in the theory of relativity, which in its early stages was to a significant degree based on the principle of observability.

Time in mythology, however, also has other aspects. For example, myth is characteristically oriented toward the past. By this is meant not the simple past of yesterday or last summer but the distant mythological past, the time in which mythical ancestors accomplished their deeds. This mythical past precedes the ordinary time of primitive mankind and its surrounding world. The mythical world is located in a time before the beginning of time. On closer examination, of course, this mythical universe proves to be not so distant after all, separated from the present by a mere four or five generations. From this it can be concluded that primitive peoples lack distant temporal perspective and that in many respects time and the past are based on memory, because the dead were remembered by children and grandchildren to whom they could appear in dreams. Only vague stories about prominent persons (who by that time had moreover been incarnated as representatives of the living) were handed down to subsequent generations; stretching even farther back is the world of mystical ancestors, which is qualitatively different and not even located in ordinary time. But mythological time has still more features: Despite its bond with the past, mythical pretime was also the present and even the future. Such a statement strikes us as extremely illogical, but the contradiction is only an apparent one; modern science proceeds from a model of linear time, whereas primitive notions are based on a cyclic model. Once this is realized, it is easy to understand A. P. Elkin when he states that "the mythical *dzugur* epoch . . . must not merely be thought of as past time, but as present and future, and a state as well as a period."[88] Events in the mythical past are an invariant structure that is simultaneous with past, present, and future. Lévi-Strauss is therefore justified in defining mythical time in terms of such dualistic features as reversibility and irreversibility, synchroneity and diachroneity.[89]

As many observers have noted, mythical time and the orientation of primitive thought toward the past have no analog in Piaget's developmental psychology. Thus Meletinskiy remarks that

a description of mythical events involving supernatural beings acting in distant times answers not so much questions [typical of the child— M. A.] as certain spiritual needs, providing a kind of "poetic" key to them. "Before" is the sphere of primary causes, the source of everything that came "after".[90]

It would be erroneous, however, to ignore totally the psychological component and to attempt to explain mythological time solely on the basis of such purely sociological phenomena as ancestor worship. It is quite true that developmental psychology does not posit a level of intelligence turned toward the past, but pathopsychology has shown that the hemispheres of the brain are responsible for different kinds of cognition and are oriented toward different times. Thus the right hemisphere controls representative, imaginal cognition and is turned toward the past. Significantly, primitive thought was in fact preconceptual, representative, imaginal, and transductive and thus dominated by right-hemisphere activity. Another reason for taking psychological features into account is that many of primitive mankind's channels of communication with the mythical past—rituals, rites, the shaman's *kamlaniye,* the fits of the sorcerer, hallucinations, actualized dreams, etc.—were clearly pathopsychological in nature.[91]

Each time a ritual was performed, a situation from the mythical past was acted out, and the present and future were perceived through the structure of the past. This factor acquired a periodic character in calendar rituals, which thereby reinforced the cyclic model of time implicitly inherent in reincarnation, even though that mechanism more nearly corresponds to a more archaic temporal model. In the matrimonial classes of the Australian aborigines, for example, the same terms are used to denote both grandfathers and grandchildren. As E. Leach observes, corresponding to this stage is an archaic sense of time as a kind of oscillation between life and death, day and night.[92] Mythology subsequently developed the notion of cyclic time, in which the present and the future are merely different temporal hypostases of the mythical past. This is the tribe's guarantee of continued existence—on the condition that mystical rites and rituals are strictly observed.

To conclude my treatment of spatial and temporal concepts in mythology, let me briefly consider the two distinctive ingredients of primitive thought. One is the logical component that underlies primitive activity and does not admit contradictions, and the other is the prelogical one that is, in effect, responsible for mythology. Introduced by Lévy-Bruhl,[93] the notion of prelogicality is manifested in primitive mankind's indifference to the contradictions typical of mythology. Note that there are two forms of this indifference, however: intermythical and intramythical. Many scholars[94] have observed that primitive culture is not particularly disturbed by the glaring contradictions that appear when myths are compared, but such intermythical inconsistency cannot be regarded as an indicator of a type of thought. If it were, modern scientists might well find themselves among the representatives of prelogical thinking, because there is no immediate logical consistency among the many physical theories (classical physics, for example, admits infinite velocity, whereas the theory of relativity sets the limit at the speed of light). The same is true of ancient myths; each of them is a completed construction rather than a chapter of a great novel in the family chronicle genre.

Certain archaic mythologies, however, are instead a collection of correlated or uncorrelated fragments. If the family chronicle develops in time, such myths are dispersed in space. Thus, as a rule among the Australian aborigines,

> no local descent group, clan or dialect unit owns a complete myth. Even though at first it may appear to do so, what it has is usually only

a section, dealing with some of the actions of a certain being. The men over in the next stretch of country may own another section, and can perform the rites associated with that—and so on, all over the country. . . . But the myth is never acted out *in toto* because all its owners could not meet, and in fact would probably not even know one another.[95]

Contradictions are inevitable in myths arising under such circumstances, but they are most probably intramythical in character.

As to intermythical inconsistencies, these are predialectical rather than prelogical. Many archaic myths arose in societies on the tribal level of organization. As psychologists such as Piaget have noted, in a number of these cultures the thinking of adults remains at the concrete operational level and never reaches the propositional stage. In other words, myths are created at the level of imaginal, transductive thought and cannot, of course, be systematized in a logically consistent form. Mankind's sensory and affective reflection of the surrounding reality was as contradictory and dialectical as the world itself. Many observers have pointed to the dual or oppositional structure of the mythological world, with its mediatory processes and intermediary links. In myth, reality is pictured rather than formalized. Moreover, myths arose in and were connected with the ritual side of society; that is, they were associated with a sphere in which dreams, ecstasy, rapture, and quite simply mental illness figured prominently. Thus ancient mythology bears the stamp not only of primitive thought but also of primitive insanity.

Another aspect pertaining to the prelogicality of primitive thinking consists of its orientation toward the mystical participation of objects and phenomena in the primitive world. Primitive culture explained all anomalies in nature or society as arising from the action of invisible, mystical, supernatural forces. A more adequate term than "prelogical" to describe this tendency, however, might be "prescientific," for modern science frequently attempts to explain anomalies by postulating the existence of hypothetical mysterious forces and objects. Thus an anomaly in the movement of Uranus was ascribed to the influence of the unknown planet Neptune, and an anomaly in β decay was attributed to the influence of the unknown neutrino particle, which quite simply possessed "supernatural" properties. Here it might be objected that such supernatural objects and forces in science will eventually be explained and included in the family of natural objects and interactions, so that it is really merely a question of the known and the not yet known. However, there is an analogous state of affairs in mythology. The mythological world is not split into natural and supernatural ingredients:[96] Primitive thought regards them as united and interpenetrating, but even if we can encounter such a doubling (it is typical of religion), mythology developed by transforming supernatural phenomena into metaphorically natural ones represented in a conventional and symbolic form. It is important to note, at any rate, that the tendency of mythology to explain anomalies as deriving from invisible forces rather than from the anomalousness of the world itself is rational and underlies the causal scientific approach to reality.

I have attempted to show that "primitive thought" is an extremely complex and multilayered notion. Describing it as prelogical states rather than solves the problem. The "pre-" indicates that primitive thought is specifically sociomorphic and refers further to its preconceptual, prescientific, dialectical, and pathological components and the means by which all these are interconnected. Naturally, this enormous question falls entirely outside the scope of the present study, but a more precise description of mythological conceptions of space and time might well contribute to its solution. For my purposes these notions have a different value, because they underlie the set of spatiotemporal structures that formed in the subsequent evolution of human cognition. Mythological space and time can be defined in the most general terms as a cyclic temporal structure and a multilayer spatial isomorphism. The structure of mythological space and time has proved heuristic. As Yu. Lotman noted: "The exceptionally strong tendency of mythological thought to establish homeo- and isomorphisms has made it scientifically fruitful and accounts for the fact that it is periodically revitalized in different historical periods."[97]

These features have both stimulated the development of myth itself and made it a fruitful source of different forms of subsequently differentiated social consciousness. Although myth canonized the mythical past, its temporal cyclicity served as an intertemporal structure that of necessity organized the past, present, and future. This in turn made it possible to reexamine the canons of the past in accordance with the changed sociocultural metrics of the present. The wheel of time rolled out of the canonized past, picked up the advances of the present, and bore them off into the past. Although the past remained canonized, it underwent changes as it accumulated the achievements of primitive thought and cognition.[98] In other words, as it rejects the archaic oscillatory model, mythological time makes the transition to a cyclic model. It does not stop there, however (the cyclic model operates in cultures that are stagnant or have an unchangeable past, that is, an immutable present and future), but goes on to develop a spiral model. This process can be traced in the cosmological and mythological conceptions reflected in the ornaments of ancient cultures. The spiral gradually supplants the circle; that is, the spiral model of time replaces the more archaic cyclic one. Two points merit attention here. First, scholars link the image of the spiral to the idea of time. Thus B. A. Rybakov interprets the typical ornamental "running spiral" of third- and fourth-century B.C. Tripolian culture as representing the course of the sun; he emphasizes that "the main idea of the Aeneolithic spiral-solar design, whose rhythmical repetition of the course of several suns consummately illustrates the continuity of that movement, is the idea of Time."[99] Spirals are found on the "cosmological" ornaments of a great many cultures of the period and are typical of this particular developmental level. Second, it is interesting to note that archeologists are presently debating the order in which the circle and the spiral appeared on ancient ornaments. I. G. Anderson, for example, has suggested that the circular motif developed from dual spiral designs, but T. I. Kashina observes that

a careful study of the extensive materials obtained in the past few years allows us to establish that the circular motif as an independent ornamental design appeared on ceramics from the Hwang Ho basin long before the use of spirals in the northwestern provinces.[100]

In later forms of social consciousness the spiral of time was straightened out into a linear construction (and here we have the appearance of history!), but mature mythology operates with the spiral model.

The idea of the Cosmos corresponds to the development of sociopolitical organization into expanding unified structures (polis, state, empire). This is manifested in late mythologies, in which primary time is occupied not by totemic animals or spirits but by cosmogony. The growth of productive forces changes mankind's view of nature, particularly the conception of the animal kingdom. Mankind begins to see itself in contrast to animals, and this strongly stimulates anthropomorphization of the gods.

The anthropomorphization of mythological figures gave rise to two currents that are diametrically opposed in their approach to the supernatural world. One of these is connected with the notion that humans can actively influence that world. Actually, in its late stages, mythology itself arrives at this idea when it acquires the character of heroic epic, in which the hero often does battle with the gods (thus Diomedes casts the god of war Ares to the ground in front of the walls of Troy and wounds Aphrodite) or defies their orders. This heroic atmosphere furthered the development of the idea that supernatural forces could be predicted, used, and even controlled; such activities became the domain of magic.

The other current also proceeds from a clear division of the world into natural and supernatural levels, but its attitude toward the supernatural is based on worship and passive faith. This is religion. N. Zederblom has ably formulated the difference between the two currents: "Magic is a matter of calculation, whereas religion is by contrast humility and submission to the mysterious and inscrutable."[101] This has led certain scholars to regard magic as prescience.[102] As for religion, its basic premises (It is absurd, therefore I believe!) have always been and will always remain antiscientific. At the same time, however, the atmosphere in which science was born and evolved was determined for many centuries by religion, particularly Christianity, and this fact obliges me to give careful attention to the spatial and temporal notions that developed in the course of religious history. In addition, it must be borne in mind that the evolution of these conceptions did not follow the linear pattern "mythology → natural philosophy → science" but included many sterile currents that nevertheless furthered ideas that considerably influenced the development of more fruitful notions.

Many contemporary theologians and philosophers of religion attempt to prove mankind's basic religiousness by equating mythology and the preanimistic concepts of religion. Despite many points of intersection, however, mythology and religion are essentially different, and the criterion for this distinction is their attitude toward time. A. Hultkrantz emphasizes that the bond mythology maintains with the distant past distinguishes it from religion, which is instead oriented toward a continuously experienced present.[103] Although one can agree with Hultkrantz that the temporal orientation of religion is different from that of mythology,[104] it is difficult to accept the latter part of his statement. The strict sacred ordering of the present, which religion does indeed demand (here we must take into account Kierkegaard's two forms of religiousness), is not an end in itself but merely a means for ensuring that certain future goals or a certain future will be realized.[105] What is the religious person's dream of the future? He hopes to free himself from time! Religion acquires its power over the souls of humans by promising them the ability to conquer time; the ancient Egyptian's blissful fields of Ialu, the Christian's Paradise, and the Buddhist's Nirvana are all located not in time but in eternity. Arising simultaneously with the world, time is the form of existence of the world between Creation and Judgment Day. For example, the Heavenly City, as described in Revelation, needs neither the sun nor the moon for light. Night will not exist there, for sacred time lacks temporal reckoning. After all, it is the sun and the moon, the alternation of day and night, on which relative time is measured in the ordinary world, whereas sacred time, the prototype of Newton's absolute time, participates in the Godhead. It is interesting to note in this connection that divine acts are thought of as instantaneous by a wide variety of religions. Already the hymns dedicated to Amon-Ra characterize him as "arriving immediately from afar to those who call upon him,"[106] and in Homer "immortal" beings are described as moving with the speed of thought. This sacred action-at-a-distance is a necessary attribute of the omnipresent and omniscient deity whose sensorium is absolute space and time.

A clear-cut distinction between time and eternity, which in science is a division into relative and absolute time (which are in turn based on the attributive and substantial notions), represents a significant advance in the development of generalizing and abstractive thought. This is amply illustrated by Feuerbach, who noted in his famous *Lectures on the Essence of Christianity*:

Eternity itself is nothing other than the concept of time, abstract time, time disjoined from temporal differences. Thus it is not to be wondered at that religion should have made time into an attribute of God or into an independent god.[107]

This subtle observation explains why time (Zurvan, kala, etc.) was regarded as the supreme deity in the ancient religions of Persia, Babylon, Phoenicia, and India. Summarizing Feuerbach's study, Lenin wrote that "time outside temporal things = God."[108] Thus evolved the notion of time as a substance existing independently of the objects, events, and processes that fill it.

One observation is in order here. It must not be thought that it was only a religious world view that brought mankind to the idea of abstract time. Even the New Testament is a relatively recent work (Revelation dates from A.D. 68–69), and it incorporates the oriental wisdom that is an organic part of the Old Testament (the Sumerian Enki and Ninhursaga, the ancient Egyptian Book of the Dead, etc.), the heroic epic, and intellectual achievements of Hellenism, such as Neoplatonism. This, of course, disrupts the linear history (if indeed such history is at all possible!) of spatial and temporal concepts. Thus, as we will see in the next chapter, certain Christian conceptions of time derive from early Greek philosophy. On the other hand, however, it would be absurd to ignore the regular process of generalization and abstraction that occurred in the course of religious history. The interesting fact here is that, as religion developed, it nourished the seed of its own destruction, for although religion is essentially antiscientific, its evolution stimulated the forms and means of thought necessary to a scientific approach to nature. One example is the way in which animism became polytheism and then monotheism. In animism each object or process had its own individual spirit or idol. The first step of generalization occurs in polytheism, in which all of nature is declassified and each sphere of reality is assigned an appropriate deity. Bodies of water (seas, rivers, lakes, etc.), for example, lose their individual gods and come under the authority of the "specific" god Poseidon or Neptune. Finally, in monotheism there is a single god with a single basic substance, and all the gods of the pantheistic pantheon are either generated by this substance or become its accidents or hypostases. This scheme can be compared with the developmental stages of physics, for example, which proceeded from the physics of properties through the physics of basic interactions to unified field theory.

There is nothing surprising in such an analogy. Religion itself is most assuredly irrational, but its metastructure, or pattern of development, is inevitably rational, reflecting the corresponding historical evolution of social being[109] and abstract, generalizing conceptual intelligence. The development of religion, therefore, can help us discern important stages in the evolution of spatial conceptions. The matter can also be stated the other way around: Spatial notions are among the most productive criteria for determining various stages or forms in the history of religion.

We can distinguish the following steps in that history, all of which are characterized by different views of space. Already during the initial stages of human history, mankind's relationship to space and territory was conditioned not only by its vital activity (gathering, commodity production, garden tillage, etc.) but also by primitive mystical or religious notions centering on death. Even Neanderthals buried their dead instead of leaving them to the mercy of fate. This testifies to the "strength and stability of the bonds uniting the primitive community,"[110] but it also tells us that humans had begun to believe in activity after death. The dead were buried and appeased out of fear that otherwise they might avenge themselves on the living. Death severed the bond of the collective with its territory. Nomad peoples even abandoned their camps to evade the malicious acts of the dead.[111] Thus a relationship between territory and the souls of the departed was recognized.

As such social structures as the clan, the community, and the family were formed and a clearer kinship system developed, mankind's attitude toward the souls of the dead also underwent a gradual change. The souls became ancestral spirits who, although they continued to be powerful forces within the clan's territory, were no longer pernicious and did not have to be avoided. In fact, on its own territory the tribe enjoyed their protection.[112] Tribal space became sacredly unique—it was the ancestral burial ground, the site of totemic centers, the locus of myths. Everything outside this space was regarded as harmful and chaotic. This affective bond between mankind and space, which might be termed *topophilia*,[113] determined behavioral norms. Feuerbach adduces some interesting facts on this subject. "The Negroes in the West Indies," he notes, "killed themselves that they might come to life again in their fatherland."[114] Thus a person might be brought to another continent, but his or her soul still belonged to the space of the tribe, whither it returned after death. Death was merely an extratemporal conveyance, a means of returning to one's native land. It is only within the territory of the tribe that the dead can inhabit a newborn tribal member; resurrection anywhere else is impossible.

This, of course, is only one way in which humans and their souls participate in tribal space. There were many such bonds, and they varied considerably among different peoples depending on their religious conceptions and level of development. In certain ancient religions, for example, the return of a soul to its native territory was thought to be impossible. Death in strange parts was viewed as a terrible tragedy, and this served to tie a people even more closely to their territory. In another type of relationship between mankind and space, not only the soul but also the *individual* was denied the possibility of returning if he was taken into slavery. The Maori, for example, considered that slaves no longer enjoyed the protection of the tribe and could not return to the tribal territory. Those who tried were sent back.[115] Here we can already see that increasing social differentiation is beginning to loosen mankind's bond to tribal space.

Subsequent socioeconomic development (the division of labor, the production and exchange of commodities, trade, cities, class society, etc.) put mankind's relationship to space in a new light. "Attachment" to one's native territory gave way to expansionism, an appetite for the territories of others, and empire building. At first the conquerors worshipped the local god of occupied lands; the space in which such a united power existed, in other words, was sacredly heterogeneous. Thus M. A. Dandamayev has shown that, at least through the reign of Darius I, the Achaemenidians adopted the Egyptian gods and worshipped Jahveh in Jerusalem, Marduk in Babylon, and the Greek gods in Asia Minor.[116] These powers, however, were unstable and were often shaken by uprisings that were put down with sadistic cruelty. The suppression of one such revolt by the Assyrian emperor, for example, is described as follows: "I beheaded their soldiers and built a pyramid of heads outside the city; I burned their children alive . . . . I impaled the surviving prisoners in a ring around the city and put out the eyes of the rest."[117] These punitive operations did not always succeed, and then the aggressors resorted to extreme measures to crush resistance. But what could be more terrible than the nightmare just described?! The extreme measure was a "spatial transformation" that assumed different forms depending on the level of religious development of the rebellious people. Three basic methods can be discerned:[118] (1) the removal of the people from sacred space, (2) the desacralization of space itself, (3) the "renormalization" of space.

The first form of spatial transformation was used when the gods were predominantly regarded as being attached to their respective spaces. In this case the people were forcefully relocated—history knows of many such examples. Gradually, however, the bond between gods and space was re-examined, a process that was facilitated by the fact that it was done first with respect to the deities of conquered nations. Here, recourse was to the second form of transformation—the god was relocated, thereby desacralizing the territory of the rebel people. Xerxes did this to punish the Babylonians: Esagila, their temple, was destroyed, and the statue of Marduk was taken to Persia (on an earlier occasion it had been removed to Assyria).

As one Russian proverb puts it, however, A holy place abhors a vacuum. Desacralized space is resacralized, but now through the god of the conquerors.[119] This is the third form of spatial transformation. In 168 B.C., for example, the officials of Antioch IV desecrated the temple in Jerusalem, destroyed Jahveh's altar, and erected another to Zeus, touching off a popular revolt against the Seleucids.

The radically new problems that these spatial transformations presented influenced human notions of space, time, and the surrounding world.[120] As gods and ancestral spirits were divorced in the first step of this process, the world came to be regarded as a number of local and finite spaces controlled by the appropriate deities. This separated the people from their tribal territory and opened up the wide world beyond its boundaries, because any travelers to new parts were obliged to worship the local gods. This shift was a response to specific sociopolitical events and processes in the ancient world, such as wars of aggression, the building of empires, migrations, the rise of trade, and cultural syncretism.[121]

The following episode related in the Old Testament is typical of this stage of religious development. The king of Assyria brought men from Babylon, Cuthah, Ava, Hamath, and Sepharvaim and moved them into the cities of Samaria in place of the children of Israel. Life in their new territory proved difficult, however, as some of them were attacked and eaten by lions. The survivors realized that the source of their woes was that they did not know the manner of the god of the land, so the king of Assyria sent them one of the expatriated priests to teach them how to worship the local god (2 Kings 17:24–28). Although the various peoples continued to erect idols and sacrifice their children to their old dieties, they were nevertheless united by worship of the god of their new land. The resulting monotheism became typical of united powers.[122] A pantheon of gods, which included representatives from the different subjugated nations and peoples, was established, usually in the capital of the empire. The space in which such empires existed was sacredly heterogeneous and anisotropic. Consolidating the state, however, naturally required a homogeneous space, and the move to create it is typical of monotheism. To effect this transition from the heterogeneous, anisotropic space of polytheism to the homogeneous, isotropic space of monotheism, a supreme deity was singled out, and the rest were reduced to its derivations or hypostases. Buddhism, for example, included in its pantheon the gods of the earlier Indian religions, all of whom became Buddha's followers and worshippers. In Hinduism, by contrast, Krishna incarnates all other gods, so that all sacrifices made to them are actually to him alone. These two forms of monotheistic religion in India are connected with various state formations and ideologies:

If the edicts of Ashoka, with whose name the spread of Buddhism is associated, did not reflect the divine origin of the royal power (a notion that was quite in keeping with the essence of Buddhist doctrine and the religious tolerance observed by the third of the Mauryas), the Gupta emperors—passionate champions of Hinduism—regarded themselves as earthly incarnations of the supreme god Vishnu.[123]

Between polytheism and monotheism there is an entire spectrum of intermediate forms. In Egypt and Kush, for example, Amon was the supreme deity, but each city had its own Amon: Amon of Napata, Amon of Thebes, etc.[124] Here there is a kind of "amonotheistic principle of relativity" that describes one transitional form between polytheism and monotheism. Polytheism posits many different gods, each of which presides over a certain territory or field of reality; in this case space is sacredly heterogeneous and anisotropic. Even the gods of ancient Greece were scattered (Zeus on Olympus, Hera in Argos, Aphrodite in Paphos, Athena in the home of Erechtheus, and so on) and did not "assemble" on Olympus until later. Furthermore, there is a "relative monotheism" in which the single god is a local deity in different territories. Although this is monotheism on the surface, space is still heterogeneous. Finally, we have "centralized monotheism," which presumes a unified space under the dominion of a single god. Thus the power of the Judaic god extended over several lands. For example, He came to the aid of His followers who had been thrown into a fiery furnace in Babylon for refusing to worship the local gods (Daniel 3:1–29). In this case, however, there exists a center from which divine power is radiated out into the world—here it is the temple in Jerusalem, to which believers far from home turn to pray.[125] Only in true monotheism (Christianity, for example) do we find a sacredly homogeneous and isotropic space that is the dominion of a single deity. In effect, such sacral space is the sensorium of the almighty god.

If in relative or centralized monotheism we can speak of finite space (the space of a nation or nations), then in true (absolute) monotheism we encounter a universal god whose space is infinite. The notion of an omnipotent and omnipresent god is incompatible with spatial finiteness, because, as Feuerbach observed, "finiteness and nothingness are identical; finiteness is only a euphemism for nothingness."[126] Consequently, the Christian world image is split into sacral and secular, absolute and relative. The centration that the Christian Cosmos possesses is geocentric in nature. Christian theology dogmatized Peripatetic philosophy and went on to dogmatize the Aristotelian-Ptolemaic Cosmos, especially because this model was in keeping with the anthropomorphic concept that posits mankind as the center and ultimate goal of the universe. But it is merely a theological model of the secular world in which time is measured by the rising and setting of the sun. It is a world in time. The sacral world, on the other hand, is synonymous with God, who in His omnipresence exists in the homogeneous, isotropic, infinite space and infinite and unmetrized time of eternity.

I have already noted that religion bears within it the seed of its own destruction. In this sense the world image that evolved in Christianity is the terminal point of religious development. The sacral view of the world gave rise to the pantheism of such thinkers as Johannes Erigena, Nicholas of Cusa, Giordano Bruno, and Spinoza, leading not only to materialism but even to certain points of contact with atheism.[127] This, of course, contributed to the development of a rational cognitive approach to nature that, in connection with the Copernican revolution, immediately exposed the flimsiness of the theological picture of the secular world. Before considering these typical problems of the modern age, however, we must first analyze the spatiotemporal conceptions of ancient Greek philosophy, whose idealist currents "led to the idea of a single god"[128] and whose materialist tendencies underlie such constructs of rational philosophy and science as atomism.

2

The Philosophical Evolution of Spatial and Temporal Conceptions

Space and Time in Early Greek Philosophy

I begin our study of spatial and temporal conceptions in early Greek philosophy with a look at these notions in mythology and the heroic epic. I start there because Homer and Hesiod were not only the precursors of the early natural philosophers but also the mentors of the entire Hellenic world.[1] As A. F. Losev puts it: "All antiquity was erected on Homer's world view and mythology."[2] As for Hesiod, modern scholars increasingly tend to regard him as the first philosopher and view his *Theogony* as a primarily philosophical work.[3] I will not repeat the material of the preceding chapter in the discussion here because in its mature (Olympian) phase early Greek mythology is a distinctive construction shaped by the wave of conceptualization that swept over the developing abstract thought of the period.[4] Of particular interest to us, for example, is the conclusion drawn by B. Hellwig in her special study of the Homeric epic that the categories of time and space were first discovered in the *Odyssey*.[5] Thus Greek mythology discovers not just any categories but the most basic ones.

The mythology and heroic epic of ancient Greece possess a number of features that are not found in archaic myths. Among these properties or features are epic historicism, the arrangement of events in a temporal sequence, and an irreverent attitude toward ancestors. Epic historicism affects both people's lives (the abduction of Helen, preparations for war, the years of the Trojan war, etc.) and divine events, as can be seen in Hesiod's *Theogony*:

Verily at the first Chaos came to be, but next wide-bosomed Earth, the ever-sure foundation of all the deathless ones who hold the peaks of snowy Olympus, and dim Tartarus in the depth of the wide-pathed Earth and Eros. . . . From Chaos came forth Erebus and black night; but of Night were born Aether and Day, whom she conceived and bore from union in love with Erebus. And Earth first bore starry Heaven, equal to herself, to cover her on every side.[6]

This union of Earth and Heaven was fertile indeed, for out of it arose the first history of the gods with a clearly marked notion of progress in time; each subsequent generation of gods creates a world that more and more resembles Cosmos. True, this gradual perfection is achieved through a process of constant generational conflict, which conforms to the following pattern: With the help of his mother (Earth) one of the sons defeats his father (Uranus is beaten by Kronos and Kronos is overthrown in turn by Zeus), castrates him, exiles him to Tartarus, and marries his mother (Earth in her various hypostases—Gaea, Rhea, Hera). This generates an increasingly Cosmos-like world (and a vast Freudian literature on mythology). This irreverent attitude toward ancestors (so unlike archaic mythology and so typical of mankind's changing relationship to the past) derives, first, from the fact that the various monsters, such as the Cyclopes and the Hundred-handed Giants that were wantonly and chaotically spawned by the first gods, were not in keeping with the Greek aesthetic sensibility, which aspired to discover the limit of the indefinite and the infinite and to mold Chaos into regular and harmonious Cosmos. Second, the ancestral gods treated their children, the gods of the next generation, with unacceptable brutality. Kronos, for example, simply ate all his offspring, and the only factor that may to some extent absolve him in our eyes is that his behavior "is the first example of the dynamic conception of time that views it as a universal process by which objects are generated and destroyed."[7]

This is all pure mythology. However, as Losev has observed:

There is already an element of gradual development in it, since "history" proceeds step by step from the chaotic fertility of Uranus to the order of reason and will which Zeus and his Olympian allies established throughout the world.[8]

This is not the only or most important feature of Hesiod's *Theogony*, however. There for the first time we encounter theology and natural philosophy, theology evolving in mythoreligious images, natural philosophy in abstract notions. As George Thomson emphasizes, Chaos and Darkness are impersonal, abstract concepts.[9] The quoted passage from Hesiod states that Chaos arose first and that after but not out of it came Gaea—Earth. Subsequent development follows two parallel courses: (1) Chaos → Night + Erebus → Aether + Hemera → . . .; (2) Gaea → Uranus → . . . → Kronos → . . . → Zeus → . . . . The second course is characteristic of mythology containing elements of natural philosophy, whereas the first marks the beginning of natural philosophy. Although it includes numerous mythological elements, the figures themselves "are not gods who once were the objects of cult worship, but personifications of cosmic notions."[10] Hesiod's Chaos merits closer attention, for it was this path that brought Greek thought to the Milesian natural philosophy of Thales, Anaximander, and Anaximenes.

As F. Kh. Kessidi observes: "Chaos is the vaguest entity in Hesiod's *Theogony*. What substance (or substances) it consists of and whether it is uniform or a mixture of substances are open questions."[11] The whole paradox, however, lies in the fact that any attempt Hesiod might have made to answer Kessidi's questions would have deflected him from the answers, because he would have been describing not Chaos but the initial features of Cosmos and its creation. Chaos is indefiniteness, boundlessness, and negativity. Chaos is darkness, a dark, yawning abyss. To the mythical consciousness such a definition would seem exhaustive. It cannot satisfy us, however, just as it could not satisfy the philosophers of antiquity. They defined the essence of Chaos within the framework of natural philosophy, and it began to figure prominently in their systems. In natural philosophy Chaos means infinite space. It would be a considerable modernization, of course, to identify Hesiod's Chaos (the word itself is generally translated as "yawning abyss") with empty space, as that notion did not appear until much later. Chaos is an indefinite receptacle that did not lack substance and materiality, and this explains how it was included in ancient Greek natural philosophy as both a primary potentiality and an empty space. Interestingly enough, modern scholars also disagree on its nature. Thus if Rozhanskiy specifically states that Earth appeared after but not out of it,[12] Kessidi writes that Earth emerged from Chaos.[13] Projected onto modern physics, these two conceptions result in different theories. In the first case we have Newton's world, in which space exists first, as a receptacle into which matter is subsequently introduced; and in the second case we have the world of Einstein and Wheeler, in which space exists first and subsequently generates the entire diversity of the material universe.

Although Hesiod's Chaos did not give rise to Earth, this does not mean that it was not a generative principle, for it did produce Night and Erebus. Thus Chaos can be defined as generative space that logically must be connected with or even contain time. Significantly, already in Pherekydes the role of Hesiod's Chaos is played by Kronos, who symbolizes both space and time.[14] The subsequent history of indefinite generative space (the substantial-material conception), however, was bound up with the fact that this space was split into its constituent components of generative principle, or primary potentiality, and empty space (the substantial-immaterial conception). The divorce was typical of various currents of early Greek natural philosophy. I consider this transformation shortly, but first let me simply note that in its syncretic form indefinite generative space retained its significance in later periods and has enjoyed a renaissance in contemporary quantum geometrodynamics. According to this theory, the space underlying the entire universe can be defined as a foamlike manifold with a fluctuating topology.[15]

Although Hesiod's natural-philosophic ideas are expounded within a theogonic framework, they use abstract notions and are much in keeping with the Greek intellectual atmosphere in the late seventh and early sixth centuries B.C. Slave-owning democracy was in its prime then and had begun to be governed by the principle that it was the duty of men to submit "to the sovereignty of the law and the instruction of reason."[16] The art of oratory and debate advanced rapidly, and the laws of thought and epistemology received considerable attention. The ancient Greek philosophers turned to the Orient, to the mathematical and cosmological doctrines of Media, Babylon, and Egypt, and to the conceptions of natural philosophy implicitly or explicitly present in the mythology of Homer and Hesiod. Natural philosophy "welded together" the intellectual achievements of preceding ages and thinkers. The product of this fruitful process was the Milesian school of Thales, Anaximander, and Anaximenes.

These thinkers were united by their search for the fundamental principle of the world, which all of them postulated as something indefinite. This did not prevent them from selecting specific phenomena, such as water (Thales) or air (Anaximenes), but the essence of these lay in their amorphousness and unorderedness. Not only the materialistic but also the dialectical tendency in the notions of the Milesians is in evidence here. This accent on the necessary indefiniteness and boundlessness of the fundamental principle implicitly present in Thales' water and Anaximenes' air becomes clearly apparent in Anaximander, who declared his principle to be *apeiron*—the purest form of boundlessness, amorphousness, and indefiniteness.[17] These Milesian conceptions echo Homer's Okeanus and Hesiod's Chaos, but they also contain a basically new ingredient that renders them distinctly philosophical: They introduced into human culture *substance,* an idea that figures prominently in modern science as well.[18] In mythology the principle is merely genetic, whereas the Milesians regarded it as substantial in nature (although it does still possess certain atavistic genetic properties).[19] Aristotle describes these conceptions as follows:

Most of the earliest philosophers conceived only of material principles as underlying all things. That of which all things consist, from which they first come and into which on their destruction they are ultimately resolved, of which the essence persists although modified by its affections—this, they say, is an element and principle of existing things. Hence they believe that nothing is either generated or destroyed, since this kind of primary entity (physis) always persists.[20]

The basic principle, then, is regarded as a kind of substance that is concealed by the unstable phenomena of the surrounding world.

Natural philosophy rejected not only the anthropomorphic gods of mythology but also the anthropomorphic genesis of objects that had been so important in theogony. It was at this point that the transition was made from the genetic to the substantial point of view, and this in turn brought about a significant development in temporal conceptions.

In mythology and theogony the fundamental principles themselves arise at a given moment in time. They are older than the other components of the world, but, like Hesiod's Chaos, they have their day of birth. True, there are certain subtleties here. Thus Chaos arose first, but time (Kronos) arises later at a certain stage in the creation of Cosmos. In other words, although the chaotic state of the world arises in the form of an initial indefinite and infinite space, its birth and chronology are reckoned in a different time from that associated with generation and destruction.[21] There is no such cyclic time in the chaotic world; time appears in a universe that is relatively but not fully ordered. There is no place for it there either, so it descends into Tartarus, which is a distant and imperfectly ordered province of the ancient universe.

It is important to emphasize that already Hesiod's *Theogony* contains, albeit not clearly, the notion of two times. One is the cyclic time of the imperfect world (Kronos), and the other is a higher time, which is used to measure events of global significance in world history. These include, for example, the sequence in which such basic principles as first Chaos and then Gaea appeared. This chronology continues until the appearance of Kronos. Ancient Greek natural philosophy developed these vague ideas of two times in terms of dynamic and static time—the time of the imperfect world of opinions and the eternity of the rationally cognizable world.

From the outset, Milesian philosophy was based on the notions that space[22] and time are eternal and that eternal movement is the source of interconversions between natural objects. In contrast to the mythological Okeanus and Chaos, the fundamental principles of the Milesians do not arise in time but are eternal. Thus Anaximander expressly states that *apeiron* is intransitory, immortal, and indestructible. Aristotle, as if attempting to explain Anaximander's conceptions, wrote that *apeiron* is deathless and imperishable.[23] Hippolyte goes even further and declares that "*apeiron* exists outside of time."[24] It is time that determines the boundaries of birth, existence, and destruction. Time appears in two projections here: A view of it from the temporal world yields the model of a dynamic time that flows from the past through the present and into the future, determining the sequence of comings-into-being and destructions, whereas viewing time from the extratemporal world results in a "spacelike" model in which past, present, and future can be taken in in a single glance. In fact, such a division becomes meaningless in the static model, because the world appears at once in its totality.

In a world operating with such temporal conceptions, necessity was naturally significant, serving as the absolute law of events. Fate did indeed play an enormous role in the culture of antiquity,[25] and although it underwent important changes in the history of ancient Greek philosophy, its first manifestation was as Necessity, or *ananke*. Fate, or necessity, and time were two sides of the same coin; together they determined the organization of the changing world.

Already in Thales we find the following two interrelated theses: "Most powerful of all is necessity, for it prevails over everything," and "Wisest of all is time, for it reveals everything,"[26] These tenets, in combination with the unity of the dynamic and static aspects of time, give us the following picture of the early pre-Socratic world. Although the future was no less real than the present, it lay ahead and was revealed to mankind only in the course of time, whereas the gods could see it at that moment. Being situated in time, humans could not see beyond the temporal horizon of the present, but from their vantage point in eternity "above" and outside of time, the immortal gods could view all of time at once. In a state of trance or ecstasy, the soothsayers, oracles, Pythians, and mantics that were a permanent feature throughout the history of ancient Greek culture could gain a glimpse into the future. The world of the future, in other words, was precisely located in time and accessible to madness.[27] Gods or oracles could reveal to the individual what tomorrow had in store, but this changed little in the order of things, because everything was determined by the iron laws of necessity. The temporal order remained immutable. Because it enjoyed the protection of Fate, anyone violating it—even the sun itself—was subject to punishment by the Erinyes. Achilles, for example, learns that death awaits him in the impending battle, but he makes no effort to alter anything. He does not desert but strides boldly into combat, filled with an unquenchable thirst for his enemies' blood yet understanding that nothing can be changed, that his death will not only occur, it in fact already *is*—not today but tomorrow.

These conceptions of time and fate characteristic of the epic and tragedy were adopted by ancient Greek philosophy, which through a rational process of abstraction separated time from the events filling it. At first there was no such division—Chaos appeared in the unity of spatial and material principles, and Kronos appeared in the unity of temporal and material principles. These components were subsequently differentiated, so that in the mature natural philosophy of antiquity space and time are separate, as are the objects and processes that fill them and are set in motion and ordered by means of specially introduced organizing principles, such as Nous and Logos. From fateful destiny Necessity becomes a law—*ananke* gives way to *nomos*—and attitudes toward the future change accordingly. Now the future could be not only "probed" through the unintelligible mutterings of the oracle but also measured with precision. Notions such as laws, Cosmos, Harmony, and Logos become prominent. All this is typical of a new phase in the history of Greek philosophy, a stage that was intensely concerned with spatial and temporal questions. As abstract, conceptual thought steadily matured, many aspects of these problems became differentiated, appearing first in a syncretic form and later generating a diversity of notions of space, time, and motion that entered into keen competition with each other.

Greek thinkers became increasingly interested in mathematical and physical problems. One such question, for example, concerned the divisibility of matter and space. Thus Anaxagoras introduced the concept of infinite divisibility into natural philosophy: "In the small nothing is smallest, there is always something smaller."[28] A spatial extension can be divided infinitely without ever reaching zero, or the mathematical point. This thesis was presented on the physical level, but it later served as the basis of early Greek continuous mathematics and the foundation of the continuous doctrine of space and time. Anaxagoras's concept of homoeomeras and their infinite divisibility in fact posed the question of unfixed infinitesimals. This idea was criticized by the Eleatic school and later developed along two lines into strict continualism and mathematical atomism. Atomism proceeded from the fixed infinitely small, such as the Democritian *amer,* and has earlier sources in Pythagoreanism.

The Pythagoreans thought of Cosmos as a harmonious unity of such basic opposites as the limit and the unlimited (that is, *apeiron*), behind which lay the contradictory unity of number and matter. A. N. Chanyshev provides an important and more precise definition that deserves to be quoted here. As the Pythagoreans understood it, matter is empty space (here *apeiron* begins to be transformed into the void): "Thus by virtue of number, Cosmos is a definite space, the unity of limit and the unlimited."[29]

It should be noted of the Pythagoreans' doctrine that, although it was formally arithmetical and geometric (atoms as elements or points), it nevertheless replaced physical atomism, because numbers were the principle of all existence, the basis of the real world: "They assumed the elements of numbers to be the elements of everything, and the whole universe to be a proportion [harmony] or number."[30] Here, as in the case of Anaxagoras, physics and mathematics and consequently matter and space have not yet been strictly separated. As for the mathematization of reality, it served a positive function because it became the basis of rational cognition. Philolaus wrote of the Pythagorean epistemology that "everything knowable has a number, for without it nothing can be understood or known."[31]

The Pythagoreans represented numbers or units by means of points, which were of course indivisible and a kind of mathematical atom. To "individualize" these units, the Pythagoreans assigned to the points square "fields" that separated them in space. When they proceeded to deal with questions of volume, the Pythagoreans operated with cubic numbers. First, this idea that the world had a discrete numerical structure served as a source of the notions of atoms and the void. Second, the Pythagoreans were harshly criticized by the Eleatics. As Theodor Gomperz observes, at any rate, Parmenides was

> acquainted with the doctrine which assumed not merely continuous space empty of all matter, but also empty interstices traversing the whole material world. The islets of matter, as we may call them, surrounded by these interstices as though by a network of canals, approximate very closely in their object and intention to the atoms of Leucippus.[32]

In this conception reality is a kind of crystal grid that, being regular in nature, serves as the basis of quantitative mathematical knowledge. Parmenides observed of this doctrine that the vacuum is "regularly distributed throughout space."[33]

This picture was discredited when it was discovered, for example, that the diagonal and the side of a square are incommensurable segments. This finding, which was obtained as a theoretical deduction not amenable to experimental confirmation, greatly upset the mathematicians of antiquity. After all, for the first time in the history of human thought the logical development of a scientific system resulted in a proposition that contradicted the fundamental thesis of the system itself. Numbers were regarded as the antipode of *apeiron* and the focus of limit and logicity, but the evolution of mathematics confronted the Pythagoreans with numbers in which there was something illogical and unlimited, numbers that contained something of *apeiron*—the famous ἄλογοι. As A. D. Aleksandrov observes:

> In the existence of incommensurable intervals the Greeks discovered a profound paradox inherent in the concept of continuity, one of the expressions of the dialectical contradiction comprised in continuity and motion.[34]

The Greek mathematicians were taken aback because the system of rational points everywhere covers exactly the numerical axis; that is, it could not accommodate what later came to be known as irrational numbers. These difficulties greatly stimulated the development of mathematics, as the problem of incommensurability generated the sophisticated theory of proportions.

The ideas of the continuous and discrete structure of space and time developed by Anaxagoras and the Pythagoreans, respectively, were examined critically by the Eleatics, who devoted considerable attention to the problem of motion. This attention was thoroughly critical, for Eleaticism was intrinsically connected with the static conception of time. They began by assuming the eternity and immutability of a single Being—there was no such thing as motion! From this position they criticized and attempted to prove the logical absurdity of rival notions. Before turning to the leading Eleatic philosophers Xenophanes, Parmenides, and Zeno, however, I must first touch on the ideas of their opponent Heraclitus, in whose dynamic concept of time the Cosmos was an ever-living fire.

Heraclitus's dynamic notion of time is connected with the idea that the world is in a process of regular evolution. "The world, which is the same for all, has not been made by any god or man; it has ever been, is now, and ever shall be an ever-living fire, kindled by measures, quenched by measures."[35] The eternal fire here is infinite time, which is divided into past, present, and future. Some scholars have taken the above proposition as evidence that Heraclitus's universe was timeless.[36] I think, however, that we can still speak of infinite time rather than timelessness, because it is meaningless to divide the latter into past, present, and future. Furthermore, the kindling and quenching of the fire corresponds to the cyclic temporal model, in which universal mutability is combined with the idea of circular motion. Each new year was evidently regarded as a repetition of the preceding one. Heraclitus establishes an organic bond between time and Logos, which is the "essence of Fate," permeates the substance of the universe, and is the "measure of time's appointed circle."[37] It is Logos that is the essence of time and that determines its basic features and structure. Yu. B. Molchanov's assertion that Heraclitus's Logos was dominated by time therefore seems rather questionable.[38] It must be borne in mind that in Heraclitus (as, incidentally, in the Pythagoreans), we encounter the origins of the notion that space and time have two aspects: On the one hand, as the void and infinite duration they are the receptacle of objects and processes; on the other hand, they are the order of these objects and processes.[39] Logos determines the universal order, the universal invariant measure of all things, the "measured" repetitive rhythm of nature. This is in fact the basis of the relational conception of space and time, with the one important distinction that in the modern notion space and time are derived as a structure of relations between material objects and processes. That is, we derive space and time as a certain order from matter in motion, whereas in early Greek philosophy this order was determined by Logos (number, Nous, etc.) and was then introduced into unordered nature, transforming it into Cosmos.

The basis of Heraclitus's world is fire, which with respect to its formlessness and indefiniteness fully corresponds to Hesiod's Chaos (and the principles of the Milesians) and embodies eternal motion. Fire is not so much a thing as a process. In this fiery world, "all things are exchanged for fire, and fire for all things, as goods are exchanged for gold and gold for goods."[40] There is something reminiscent of the Pythagorean outlook in this thesis—in the clink of gold and the crackle of the blazing fire one seems to hear numbers, quantitative relations, proportions, and the like. There is an unquestionable affinity between Pythagoreanism and Heraclitus's doctrine, although the latter contains no number mysticism. Whereas the Pythagoreans attempted to penetrate to the truth of the world by means of numbers, Heraclitus was guided instead by Logos (the "fire-Logos" link in Heraclitus corresponds to the "fire-number" of such Pythagoreans as Philolaus). Not number but the Word is the basis of the world! To be sure, Heraclitus uses peculiar language. As well he might, for he was not describing the everyday world but was trying to penetrate to the true essence of phenomena. This true essence of things could be discovered through language, which contains their secret, their Logos.

In the twentieth century, mathematics has come to be regarded as a kind of language, and from this viewpoint the Pythagorean doctrine can be considered the first attempt to know the essence and laws of the world by means other than ordinary language. Heraclitus similarly aspired to discover the dialectical essence of the world, and he as well therefore resorted to unusual language. Failure to understand this accounts for the view of Heraclitus as the "dark" or "obscure" philosopher. Some people knew Pythagoras's mathematics, so they did not regard it as "dark"; others were not familiar with it and therefore had no idea whether it was "dark" or "clear." As for Heraclitus's language, it seemed deceptively familiar to all; only his meaning was unclear, and this won him the epithet of the Dark. The crux of the matter, however, is that Heraclitus was speaking in a different language about something different. Kessidi wrote:

> Words and the notions we attach to words to designate real phenomena dismember things and phenomena, presenting the latter as static and immutable, so that a unified and living whole is broken down into a number of isolated elements. The ancient philosopher [Heraclitus—M. A.] aspired to express the opposite of this, namely the vitality and activeness of being, its wholeness and contradictory essence.[41]

Heraclitus realized that formal logical analysis could not deal with motion, and he therefore attempted a dialectical reworking of language and logic. (There is another approach, taken by the Eleatics, that proceeds from the stability and absoluteness of formal logical analysis to deny the reality of motion.) In this way Heraclitus discovered the fundamental formulas that were incorporated into the mature dialectics of later ages in the systems of Hegel, Marx, Engels, and Lenin. The following propositions can serve as illustrations: "We step and we do not step into the same river"; "we are and we are not"; "it is impossible to step twice into the same river."[42] More is at issue here than just the universal dynamism of the world that fills Heraclitus's famous "Everything flows!"—a proposition that, according to Hegel, proclaims becoming to be the basic definition of all being.[43] The problem concerns the essence of motion, and the solution to it is sought in the dialectical unity of opposites expressed by Heraclitus's favorite metaphor of the river into which we do and do not step.

The image of the river is so dialectical in nature that diametrically opposed conclusions have been drawn from the theses in which it is used. Some scholars hold that it is the apotheosis of mutability, whereas others assert that what is meant is immutability and immobility. Claude Ramnoux, for example, assumes that Heraclitus was advancing not the thesis of eternal flux but the opposite notion that existence is constant and immobile, "since there is nothing more stationary than a flowing stream,"[44] In my opinion, both extremes are equally alien to Heraclitus and blunt his dialectics. Heraclitus's notions are not the naked relativism of Kratylos and do indeed contain a certain constancy and invariance. These invariants, however, are geometric rather than dynamic, as is apparent in the thesis "change reposes."[45] But this is not all. The central point is that these two opposite tendencies are at one in Heraclitus's system, and herein is the essence of his doctrine:

⟨All is one: the divisible is the indivisible, born and unborn, mortal and immortal, Logos and eternity, father and son, God and justice⟩; heeding not me but Logos it is wise to admit that all is one.[46]

This slogan (all is one!) was most fully developed by the Eleatics, whose philosophy was explicitly anti-Heraclitic. Actually, Heraclitus was not very well disposed toward the Eleatic school either. Thus he said of its founder, Xenophanes, that "much learning does not teach understanding."[47] The source of his attitude may lie in the fact that although Xenophanes possessed enough knowledge to comprehend the thesis that all is one, he chose another, anti-Heraclitic course and announced that the One is immobile. Such a pronouncement could not strike Heraclitus as wise. But the history of philosophy passed its own verdict: Xenophanes was intelligent enough to found one of the most interesting philosophical schools of ancient Greece, a school that not only developed original notions of being, thought, motion, space, and time but also subjected all previous philosophy on these topics to a penetrating criticism.

The singular way that Xenophanes arrived at his idea of the One shows that, besides critically reviewing mythology and struggling with the naive mythological outlook, Greek philosophy also criticized religion and opposed the primitive world view. To his notion of the One, Xenophanes brought a critical analysis of the extant ideas of the gods. During his many travels he had noted the curious fact that different gods resembled the peoples that worshipped them. The Ethiopians said that the gods were black and snub-nosed, whereas the Thracians claimed that their gods had blue eyes and red hair. This led Xenophanes to the blasphemous conclusion that

> if oxen, horses and lions had hands and could paint and produce works of art as men do, horses would paint the forms of gods like horses, oxen like oxen, and make their bodies in the image of their several kinds.[48]

All this suggested to Xenophanes that the one God neither looks nor thinks like mortals. Xenophanes' God is more pantheistic than monotheistic; He is the One Being. God, that is, the One, is spherical, homogeneous, eternal (in certain versions eternity itself), and "He with the whole of his being beholdeth and maketh and heareth."[49] As Simplicius tells us: "He remains always in the same place, moving nowhere; it does not befit him to move from place to place."[50] Indeed, why should He move? Motion, after all, is the subjugation of space and time, but these have already been conquered from the beginning by the One, who is omnipresent and who sees, thinks, and hears with all His being. Here we have a conceptual forerunner of Newton's action-at-a-distance in the absolute space and time of the *sensorium Dei*. Xenophanes' One is alien to the attributive notion of space and time, being a modified substantial conception that views them as a homogeneous spherical volume and a pure, limitless duration sublimated in eternity.[51] To be sure, the One is a closed sphere, but this closedness should be regarded as axiological rather than spatial, because the sphere was held to be a perfect figure. Melissus, who was more of a physicalist, later reviewed this thesis and introduced the notion of infinite space into Eleaticism.

As we analyze the conceptions that brought Xenophanes to the idea of the One, we must keep one important fact in mind—the question that exercised him was how to reflect something higher, divine, and immortal in the ordinary concepts developed in the world of mortals. Can the notions of the world of opinions express anything from the world of truth, or must we who merely possess opinionative intelligence always endow the gods with our own shortcomings? Can the substances, properties, and conceptions characteristic of mankind and its changing opinions participate in a higher being? Heraclitus tried to solve this problem by describing the essence of substance and being as a unity of opposites. That is, if in the everyday world there is either limit or the unlimited, either motion or rest, in the world of substance the two are united, as in the unity of limit and the unlimited. Xenophanes' approach was different but also dialectical. He concluded that neither one nor the other is applicable to substance if this "one" and "the other" are taken from the ordinary world. Strictly, the One of Xenophanes is not simply motionless; it contains neither motion nor rest! Theophrastus provides a good account of this feature of Xenophanes' conceptions:

It [the One of Xenophanes—M.A.] is neither limitless nor limited, since on the one hand the unlimited is not-being, having no beginning, middle nor end, and on the other hand, only the many can limit each other. He denies motion and rest in precisely the same way. . . . By the presence of being he means not a state of rest as opposed to motion, but a state that is alien to both rest and motion.[52]

Thus Xenophanes espoused the agnostic view that the truth is inaccessible and that, even if someone should accidentally happen to utter it, he and those around him would not understand it, owing to the mass of untrue theses under which they were buried.

No man has seen what is clear nor ever will any man know it. Nay, for e'en should he chance to affirm what is really existent, He himself knoweth not, for all is swayed by opining.[53]

Xenophanes' doctrine of the One was accepted and elaborated by his follower Parmenides, who began by refuting his teacher's agnosticism. If opinionative reason is the criterion of truth, then such truth is worthless, argued Parmenides:

He rejected the opinionative reason—I mean that which has weak conceptions—and assumed as criterion the cognitive—that is the inerrant—reason as he also gave up the belief in senses.[54]

Accordingly, Parmenides distinguished two types of knowledge corresponding to the two levels of the world: There was the immutable, uncognizable world of the One and the changeable world of plurality, truth, and opinions accessible to the senses. Because reason is taken as the criterion of truth, Parmenides based his doctrine on rigorous logic. This applies not only to the doctrine of the One itself but also to his proofs of the contradictions and falseness in the doctrines that assume the multiplicity and dynamism of the world. Zeno especially had advanced the notion of dynamism. All these problems are intimately connected with conceptions of space and time.

The teaching of Parmenides is predicated on the thesis that "all reality is one." It follows that temporal sequence is eliminated and replaced by eternity. There was no place in the conception of One and indivisible being for the void, which was regarded as not-being and the necessary condition of motion. If Xenophanes held that humans have only opinionative reason and cannot think of or imagine the truth of being, Parmenides is of the opposite view that the truth of being is accessible to reason and that it is therefore impossible to conceive of not-being. This served to confirm the nonexistence of not-being, that is, empty space, and implied that the One had nothing to do with space and motion.[55] Parmenides presents these ideas in his poem "On Nature":

What is is unborn and imperishable; whole and unique, and immovable, and without end (in time); nor was it ever, nor will it be, since it is now all at once, one, continuous. For what birth of it wilt thou look for?[56]

Here Parmenides divides time into past, present, and future, although he notes that it is meaningless to do so in the world of the One. Such a division operates in the world of arising and perishing objects that is accessible to the senses. As for the One, it is entirely in the present. Commenting on this thought, Molchanov writes that only the present moment possesses being and that in this sense it is identical with eternity and timelessness.[57]

As to the logical critique of rival doctrines, this is basically associated with Zeno, whose entire thought was devoted to proving the logical contradictions in ideas of plurality and change. In "On Nature" he wrote that, if reality is plural, it is both great and small—so great that its size is infinite, and so small that it has no size at all.[58] He adduces this proposition to expose the weakness of conceptions founded on infinite divisibility or points. Zeno demonstrated that these notions could not construct a world of finite objects because the only possibilities they allowed were either zero or infinity. Here we encounter the dilemma on which the aporias of motion are based: Either finite objects have no extension, in which case even adding them together will not produce a magnitude, or they have at least a tiny magnitude, in which case an infinite number of them will result in an infinite magnitude.

Certain conceptions of plurality proceeded from a recognition of empty space. Zeno attempted to prove the absurdity of such an assumption. If all that exists exists in space and space itself exists, then where does space exist? In some new space? But then a similar question presents itself with respect to this new space, and so on ad infinitum. Consequently, empty space does not exist.

Zeno exposed the logical contradictions of rival conceptions in his famous aporias, which continue to be relevant even today. Of the enormous number he devised, only a few have survived, the best known being "The Dichotomy," "Achilles," "The Stadium," and "The Arrow."

In "The Dichotomy" the nonexistence of motion follows from the proposition that, before a moving body can traverse a distance, it must first traverse half that distance. If the half is divided into a quarter and so on to infinity, one reaches the conclusion that the body cannot even start moving, and motion is therefore impossible. "Intimidated" by the insolubility of infinite regression, the character in "The Dichotomy" cannot stir from the spot. There is an analogous argument in "Achilles," in which the fleet-footed hero surrenders in the face of infinite divisibility and admits he will never overtake the tortoise:

The slowest runner will never be overtaken by the swiftest, since the pursuer must first reach the point from which the pursued started, and so the slower must always be ahead.[59]

Zeno was attempting to show in these paradoxes that infinite divisibility is incompatible with motion and leads to illogical conclusions. G. J. Whitrow observes quite correctly that

Zeno's paradox is not concerned with the question of whether Achilles does catch the tortoise but with the application to the study of motion of the hypothesis of the infinite divisibility of space and time.[60]

In "The Stadium" Zeno tried to demonstrate that a sequence of indivisible instants (of minimal duration) cannot exist, that time is infinitely divisible.[61] As Whitrow once again correctly notes, however, even though "The Stadium" is based on the hypothesis of spatial and temporal atomicity, Zeno implicitly appeals to continuity in his discussion of it.[62]

The essence of "The Arrow" is the assertion that an arrow in flight is at rest. Aristotle pointed out that this conclusion follows if time is conceived as being composed of individual "nows"; if this is not admitted, there is no syllogism. In this aporia the nonexistence of motion is demonstrated within the framework of conceptions of points. Such notions regard motion as the sum of states of rest, because at a given "now" the arrow is at some given point in space (its position being determined by some single point on it such as the tip), at another instant it is at a different point, and so on. On the basis of such an abstract mathematical approach, it is impossible to perceive the transition from one point to another. "The Arrow" is sometimes regarded as a reaction to the Pythagorean doctrine. This may be correct with respect to the history of the problem, but in terms of logic it does not quite fit the facts because the trajectory of the arrow Zeno is operating with is a continuous line consisting of an infinite, unbroken number of positional points. The Pythagoreans, on the other hand, envisage the line as consisting of a finite, that is, countable, number of points or units. The main target in this case was in all probability Heraclitus's dynamism—the thesis "that which moves is at rest" is set in opposition to "everything flows." Paradoxical as it may seem, however, both Zeno and Heraclitus come to the same conclusion and formulate the same antinomy. Heraclitus's paradox can be stated as follows: "You cannot step twice into the same river; for fresh waters are ever flowing in upon you." Zeno's is: "A flying arrow is at rest at each instant: at time t_1 it is at point x_1, at time t_2 it is at point x_2, etc." But this latter paradox can also be rephrased to read: "A flying arrow cannot be at the same point twice." In that case it becomes apparent that we have to do with one and the same paradox—once with reference to the position of a point at rest in a moving stream and then with respect to the position of a static point of the stream or "flight" itself, that is, the flying arrow.

It is no coincidence—indeed, it is inevitable—that both the

dynamist Heraclitus and the staticist Zeno should have arrived at the same paradox. Heraclitus consciously sought the essence of being in the unity of opposites, declaring when he had achieved his goal that "it is in changing that things find repose"; Zeno attempted to refute his rivals' conceptions strictly logically by exposing their contradictions, and he considered he had succeeded when he could deduce the thesis "rest is in flight" from his opponents' constructs. To refute dynamism with contradictions, however, is merely to add fuel to the fire.

The state of affairs described above is a common one in ancient Greek philosophy. Thinkers of different metaphysical schools often draw diametrically opposed conclusions from the same propositions; incorporated into different conceptual systems, one and the same thesis confirms totally different and often opposite constructs. This seems to stake out the ground on which in the twentieth century the experiment would arise as the criterion of choice between rival theories. The Eleatics, for example, regarded the void as a necessary condition for motion and therefore denied the existence of empty space because the One is immobile. The atomists also considered the void as a necessary condition for motion, but for that same reason they refuted the immobile One of the Eleatics.

The atomism of Leucippus and Democritus evolved as a synthetic doctrine that elaborated the rational tendencies of Pythagoreanism, the plurality of Anaxagoras, and the dynamism of Heraclitus but that took into account throughout the critical arguments of the Eleatics, whose One was not rejected but merely broken down into an infinite number of fragments or atoms, which scattered into infinite empty space. These eternal fragments of being moved about forever in infinite notbeing, and (full) being did not exist to any higher degree than (empty) not-being.[63] Atoms do not arise, because time has no beginning. Aristotle emphasizes that Leucippus (like Plato) regarded eternity as actuality; in this sense motion is eternal: Atoms move in the void of infinite time.[64]

The atoms of Leucippus and Democritus are concrete products of the material analogs of the Pythagorean "forms" and "ideas" that have been endowed with the properties of the Eleatics' One. Atoms are physically indivisible because of their density and lack of emptiness. To serve as the basis of natural diversity, they must move and unite in different combinations.

Proceeding from this demand, the atomists postulated the void as a necessary condition for motion. Empty space is substantial—a substantial potentiality of motion.

As S. Ya. Lur'ye demonstrates in his interesting study, the physical and mathematical aspects in the development of Leucippus's and Democritus's atomism were united.[65] The continued failure of studies in ancient Greek philosophy to take sufficient note of this fact, however, often results in incorrect interpretations of many theses of the doctrine. Thus one encounters the strange assertions that atoms undergo qualitative changes,[66] that Leucippus's being and nothingness (that is, atoms and the void) interconvert,[67] and that the absence of empty space in the atom implies that the atom has no parts.[68] Such descriptions of atomism seem in great measure to derive from a failure to distinguish clearly the physical from the mathematical aspects.

Infinitely varied in shape, size, and order, (physically indivisible) atoms combine with empty space to form the entire content of the real world. They are made up of *amera*, which are truly indivisible, have no parts, and serve as the criterion of mathematical indivisibility. Physically indivisible atoms never break down into *amera*. *Amera* do not exist in a free state. These notions of early Greek atomism are much in keeping with the ideas of modern physics, which holds that elementary particles consist of quarks, which evidently do not exist in a free state.

The *amer* is the spatial minimum of matter, the "atom" of discrete space on which all atomistic mathematics was based. Explaining the function of *amera* in atomist philosophy, Epicurus noted: "We must consider these least indivisible points as boundary-marks, providing in themselves as primary units the measures of size for the atoms."[69] We see that Epicurus took a metric approach to *amera*: Material in nature, *amera* serve as the absolute scale for measuring extension in the atomic world and are the primary elements of atomistic geometry. These notions show the abstract mathematical and the physical approaches to be in profound dialectical unity, because

if everything physical ultimately dissolves in the mathematical (in geometric shapes and in numbers), Democritus, by contrast, regards the mathematical itself as a physical reality.[70]

Many scholars have denied the reality of *amera*.[71] Specifically, it has been maintained that Democritus only "mentally perceived" them as parts of atoms.[72] Such assertions are groundless, as can be realized by turning to the epistemology of the atomists, according to which knowledge is either illegitimate (or obscure) or legitimate (true knowledge). The relationship of the types corresponds to that between sensory and rational knowledge. Democritus held that phenomena perceived through the senses exist in and are dependent on only opinion. The only true essence is a rationally cognizable one. Bearing this in mind, it becomes apparent that by "mentally perceived" Democritus was referring to the truth and objective reality of *amera*. An extant fragment written by Democritus himself can shed more light on these words than accounts by the commentators of antiquity:

Whenever the bastard kind [of knowledge] is unable any longer to see what has become too small, or to hear or smell or taste or perceive it by touch ⟨one must have recourse to⟩ another and finer ⟨instrument⟩, since in thought it has a more delicate organ of knowing.[73]

The object of the sensible world is subject to physical fragmentation. The limit of such division is the atom, which is physically indivisible. It is at this level that the transition from obscure sensory knowledge to true rational knowledge occurs. Atoms can be known because they are subject to theoretical, mathematical division, and this reflects the presence in them of a certain structure. Losev has advanced some interesting ideas in this context on the internal structural-numerical nature of atoms.[74] Even theoretical analysis, however, has a limit beyond which are encountered qualitatively different objects. These are *amera*, which are mathematically indivisible, lack parts, and serve as the boundary of analysis and deduction. Here, of course, the question arises, How can one conceive of even a minimal extension that lacks parts and form? The extension of the *amer* has been described by Lur'ye, who notes that "this particle, if one can call it that, is . . . the pure principle of extension."[75] It should be observed that during the past two and a half thousand years little has changed in our notions of the "atoms" of space. Even today, when we operate with the concept of elementary length in physical theories, we are in fact attributing to it the same lack of parts, from which follows the absence of left and right, temporal order, cause and effect, the point location of events, etc. In reconstructing the system of Democritus as a theory of structural levels—the physical (atoms and the void) and the mathematical (*amera*)—we inevitably encounter two spaces: continuous physical space as receptacle (that is, Democritus's void) and mathematical discrete space in which *amera* serve as the standards for measuring the extension of matter. The presence of two spaces was noted already by E. Frank, who did not, however, understand their singularity and concluded that Democritus

distinguishes empty mathematical space (the ideal space of geometry) from real physical space. To mathematical space (pure "nothing") he attributes infinite divisibility, whereas physical space is not infinitely divisible and consists of discrete spatial elements.[76]

Physical and mathematical space have obviously exchanged roles in Frank's reconstruction. If we followed his logic, we would be forced to consider that Democritus's system recognizes continuous, infinitely divisible mathematical space alongside mathematical atomism. In other words, it is not the space based on indivisible mathematical *amera* that is mathematical but the void—the second principle of the physical level. This inconsistency in Frank's treatment did not escape A. O. Makovel'skiy, who writes:

Democritus does indeed attribute an atomistic structure to real space, thus denying its continuity and infinite divisibility. His entire mathematics is erected upon this atomistic understanding of space, and there are no grounds whatever for speaking as Frank does of a special mathematical space distinct from real space in his system.[77]

I share Makovel'skiy's critical impatience but would add that nonetheless we must not lose sight of Democritus's second physical space, which is no less real than the mathematical. This is the void (the second principle of the physical level), which is the continuous and unlimited arena of motion, interacting atoms, and material objects in general.

Such divisions into continuous and discrete spaces were typical of the early atomists. Thus, in ancient Indian doctrines (and not only there), we find on the one hand *akasa*—boundless space—and on the other, *dish,* the space of geometric figures in which direction and position are defined. That is, as in the case of Democritus, besides empty space there exists yet another space that admits metric relations.[78] As an example one could also cite the later atomism of Newton, whose concept of absolute and relative time and space will be treated in detail later. As for the two spaces in the system of Democritus, it must be remembered that they are both real. Democritus's physical space, however, has nothing in common with geometry. He did not ascribe any metric properties at all to the void—*apeiron,* as he called it, is not geometric.[79] Empty space is simply a necessary condition for the motion and existence of atoms. Aristotle emphasized that the void *qua* void has no differentiae. Its nonextension follows from its indifference—it is negative throughout.[80] This problem of nonextension has been ably analyzed by Lur'ye:

> Empty space (τό κενον) is τό μηὸν, "non-existent"; more precisely, being is intrinsic to it in a different sense than to matter. Extension is a category of matter; the void is non-extensive; in the void there are no . . . distances, and therefore from Democritus' point of view "a straight line drawn in space" does not exist.[81]

Failure to take these observations into account has sometimes distorted the notions of antiquity. V. Ya. Komarova, for example, considers that Democritus could not "call something that is extensive corporeal, since the void is also extensive,"[82]

In accordance with the atomism of space, Democritus assumed the atomistic nature of time and motion. As subsequently elaborated by Epicurus, these ideas were developed into an orderly system. Epicurus treated the properties of mechanical motion in the context of discrete space and time: isotachys, *kekinema,* and renovation.[83] Isotachys means that all motions occur at the same speed. Here we see a further development in the problem of the dynamics of being: If in the being of Heraclitus everything changes and in that of the Eleatics everything is unchangeable, the atomists consider that everything changes at an unchangeable rate. Isotachys is typical of both aspects of Greek atomism. Epicurus wrote Herodotus that "atoms must move with equal speed, when they are borne onwards through the void, nothing colliding with them."[84] Aristotle reasons similarly; the void offers no resistance to anything—large or small, heavy or light—and therefore everything travels with equal speed.[85] Isotachys on the level of physical atomism is the forerunner of Newton's first law of mechanics, which states that every body perseveres in its state of rest or uniform motion in a straight line unless compelled by external forces to change that state. In Epicurus's words, that is if nothing counteracts it. On the mathematical level isotachys consists of the following: In the process of movement, an object passes one "atom" of space in one "atom" of time (otherwise, as Aristotle convincingly demonstrated, the indivisible can be divided), and this of course is what provides for the existence of a basic, constant rate of motion. As Sextus Empiricus observed, when a body passes one indivisible place in one indivisible period of time, all motions are of equal velocity.[86]

The essential property of *kekinema* is that along an indivisible distance nothing can be in the process of moving but can only move.[87] There is no motion in an "atom" of space. As Alexander of Aphrodisias wrote, "On each indivisible path there is no motion, but only the result of motion."[88] Motion consists not of states of rest but of "atoms" of motion—*kinema*.

Renovation implies that a particle in elementary motion cannot pass through the "atom" of space point by point; that is, it is not at all its points, because the "atom" of space, or *amer*, has no parts. The particle moves at once over an elementary length. It disappears, as it were, and after an "atom" of time, arises in the neighboring cell, moving to an "atom" of space. Here, as in Zeno's aporias, although we speak of bodies, particles, and so on, the participants are clearly mathematical objects (*amer, kronon, kinema*, etc.). Nor is there even any question here of atoms, because they would not fit in the elementary cells of space. Moreover, in renovation moving objects constantly oscillate between being (when they are in the cell) and not-being, because the transition to the neighboring cell takes place extraspatially. The atom in the void is a model of being in not-being, but if it begins to disappear in renovational "flickers," it will leave not-being as it becomes not-being itself. Thus, in discrete time and space, the properties of mechanical motion considered here are not related to the physical level but operate within mathematical atomicity, in which *amera* are not so much components of atoms as independent essences. It was at this level that the problems posed in the aporias of the Eleatics had to be worked out. In the principles of renovation and *kekinema* the early Greek atomists attempted to penetrate the dialectics and contradictory nature of motion. Renovation is a forerunner of the modern dialectical view of motion, which addresses the question of how a body can simultaneously be and not be located at a given point in space.

Mathematical atomism and the conception of discrete space and time underwent further development and modification in the different philosophical systems of ancient Greece. From the scholia to Aristotle we learn that

> of those who have advanced the doctrine of indivisables, some, like Leucippus and Democritus, state that there exist indivisible bodies; others, such as Xenocrates, admit indivisible lines, and Plato allows for indivisible planes.[89]

In his teaching Plato attempted to synthesize the rational aspects of the pre-Socratic naturalism that culminated in atomism. The subjection of Empedocles' elements to strict mathematical proportions and relations reveals the influence of the Pythagoreans. From atomism Plato borrowed the notion of indivisible particles as the basis of the elements, and he also took the features of mathematical atomism. He linked the natural elements with regular polyhedrons borrowed from the Pythagorean Philolaus. The earth is a hexahedron (six surfaces), air is an octrahedron (eight surfaces), fire is a tetrahedron (four surfaces), and finally there is the dodecahedron (twelve surfaces), which "God used up for the Universe in his decoration thereof."[90] The surfaces of these geometric bodies consist of two types of elemental right triangles whose sides have the ratios $1:1:\sqrt{2}$ and $1:\sqrt{3}:2$. Elements with surfaces consisting of identical elemental triangles can interconvert; that is, elemental triangles are redistributed in different configurations to form different elements. Also, the triangles evidently differed in size, as one and the same element has varying dimensions depending on its type; air, for example, has several modifications, such as ether, mist, and darkness.[91]

The Platonic doctrine is curiously refracted in modern physics. Certain Western philosophers and natural scientists conclude that the elementary particles of contemporary physics are desubstantialized, asserting that the basis of all reality is not matter but mathematical shapes and regarding quantum physics as a turn from Democritus to Plato.[92] Such an approach ignores certain factors. First, it is in general incorrect to oppose Democritus's atomism to that of Plato's because Plato's atomism is an elaboration of the same doctrine. Second, it must be borne in mind that Plato's elemental triangles are not purely mathematical objects. As ancient commentators such as Aristotle and Simplicius tell us, these triangles differ from both purely mathematical shapes and objects of the physical world, being mathematical shapes that possess certain physical properties. Losev therefore regards them as a real organization of space. "They speak not of some ideally geometrical surface," he writes, "but are formulae of a space which, having all three dimensions, is organized in a certain way."[93] Moreover, the triangles are not just spatial geometric figures; they also serve to delimit space. Space is what is contained in the regular polyhedrons of the elements.[94] Here it is but a step to asserting the ideality of Plato's primary principles; what could be more ideal than empty space organized by mathematical shapes, that is, mathematically organized elements of space?

Democritus's atom is an element that contains no empty space; that, in fact, is what makes it an atom. At first glance Plato would seem to represent the opposite case: An element contains nothing but space. To understand Plato's conception, however, it must be determined what he meant by space, which is not a form of the elements but rather their substantial content. Space, or as Plato himself called it, "the nurse of all becoming," is a corporeal, material principle.[95] Being an idealist, Plato considered the Idea (world of ideas) to be identical with being, but in the aspect under discussion here the basis of everything is material space.

Plato's concepts of time are likewise not strikingly original but derive from the Eleatics and the atomists. Being has no beginning and is eternal and timeless. This rationally cognizable reality stands in contrast to the world of opinions in which time passes and the processes of coming-into-being and passing-away occur.[96] Time arose simultaneously with the heavens and is inseparably connected with them, as it is metricized by their movements. Plato regarded time as the moving image of eternity. This scheme proved to be so accurate that it is encountered much later in Newton's system, where relative time is in fact the moving empirical image of absolute time, which, under conditions of instantaneous action-at-a-distance, tends instead to acquire the character of static time or eternity. This state of affairs is still relevant today, because modern science cannot dispense with the empirical interpretation and verification of theory and because empirical time is the moving image of the theoretical structure. Plato faced a similar task. He had to establish a correspondence between the real world and the world of opinions. This was why God

> planned to make a movable image of Eternity, and, as He set in order the Heaven, of that Eternity which abides in unity He made an eternal image, moving according to number, even that which we have named Time. For simultaneously with the construction of the Heaven He contrived the production of days and nights and months and years, which existed not before the Heaven came into being. And these are all portions of Time; even as "Was" and "Shall be" are generated forms of Time, although we apply them wrongly, without noticing, to Eternal Being. For we say that it "is" or "was" or "will be", whereas, in truth of speech, "is" alone is the appropriate term; "was" and "will be" on the other hand, are terms properly applicable to the Becoming which proceeds in Time, since both of these are motions.[97]

This passage from *Timaeus* is a good description of the dialectics of eternity and time. First, Plato does not deny the reality of past or present time, but merely says that it is inappropriate to divide time into past, present, and future with respect to the world of ideas. This world possesses all time at once, in all its wholeness and indivisibility; this world always "is," existing in eternity. Second, he clearly indicates the numerical nature of time, which is metrized through the revolution of the heavens—time is ordered by the laws of the Cosmos.[98] Plato's time and eternity do not coincide if only for the reason that time is younger than eternity; yet they are also identical because time is the moving image of eternity and eternity is unmetricized time.

The notion that time is numerical in nature and metricized by the motion of the celestial bodies was developed further in the system of Plato's great follower, Aristotle, who was in general critical of his teacher's philosophy. Aristotle does not present a complete doctrine of space and time all at once but gradually unfolds the step-by-step process by which the essence of these basic categories comes to be known. He first asks whether time exists at all. Then, taking an abstract mathematical approach on the level of Zeno's aporias, he adduces evidence in support of the view that it does not exist (or at best, "hardly exists").

Indeed, the past is no longer, the future is not yet, and there is only a durationless "now" devoid of length and squeezed between the nonexistent past and future. But that which is composed of something nonexistent cannot participate in being.

In addition, if anything with parts is to exist, then, when it exists, all or some of its parts must exist. But, although time is divisible, some parts of it have been and the others will be, and no part of it exists. And as for a moment, it is no part of time; for a part measures the whole, and the whole must be composed of the parts, but it is thought that time is not composed of moments.[99]

Thus Aristotle begins with the general question of whether time exists and then transforms it into the question of whether divisible time exists. Here he is incidentally polemizing with philosophical currents that regarded time as divisible and composed of "nows." His argument was probably addressed to the atomists, who allowed for the discreteness of time and space and admitted renovational motion. To Aristotle the moment, or "now," was not simply an element of discontinuity but was to a greater extent a linking element that continualized temporal duration: Time is both continuous "by means of a moment and divisible with respect to a moment."[100] Between any "nows" lies duration, just as a line lies between points. Here Aristotle allows that "now" is not always uniform and identical, that it differs from purely mathematical objects such as the point. As for his analysis of time, it was conducted in an abstract mathematical key; there is no place for motion, and this determines time to be static. On this level of his temporal analysis Aristotle concludes that space, time, and motion are continuous, "for neither is time composed of moments nor a line of points nor a motion of [indivisible] impulses."[101] In regarding motion as a point correspondence of a given place occupied by a moving body to a given "now," however, Aristotle perceives another factor, namely, that in the indivisible "now" there is neither motion nor rest. If there were motion, for different velocities one would have to divide time differently for one and the same spatial magnitude, but "now" is an indivisible magnitude. As for rest, it would exist if a body were located at a given point "now" and before, which does not happen in the case of a moving body. Although Aristotle did not actually arrive at an abstract mathematical expression of time and motion, he is here pointing to an escape from its confining limits; in this sense he is methodologically more advanced than the natural sciences of the seventeenth and eighteenth centuries.

Aristotle's further analysis of time is on the physical level, most of his attention being concentrated on the relationship between time and motion. He demonstrates that, although time is inconceivable and nonexistent apart from motion, it is not identical with motion. He specifies what motion he means:

Some say that time is the motion of the whole (Universe); others say that it is the sphere itself. But then a part of a revolution will be time, and it ⟨i.e., that part⟩ is certainly not a revolution; . . . Moreover, if there were more than one heaven, the motion of any one of them, like that of any other, would be time in a similar way, and so a plurality of times would exist simultaneously.[102]

The motion of the celestial bodies determines the periodical process by which the flow of time is metricized. Time can be metricized through any motion, but if the physical magnitude obtained is to be universal, one must use a motion of maximum velocity. In modern physics this velocity is the speed of light, whereas in ancient and medieval philosophy it was associated with the movements of the heavens. Aristotle emphasized that "they know motion by simple motion and the most rapid. . . . Hence in astronomy . . . they assume that the motion of the heavens is uniform and the most rapid, and by it judge the others."[103] Time is universal and isotachic. Whereas motion and change can be found in one place and not another, "time is equally present everywhere." Further, every change is faster or slower, but "time is not; for the slow and the fast are defined in terms of time."[104]

Universal and isotachic time is the measure of all motion and rest of objects and processes in the objective world. Some observers have regarded this as a concession to the substantial conception of time. Yu. B. Molchanov, for example, writes: "On the one hand . . . Aristotle defines time by the motion of material objects, thus denying it an autonomous status. On the other, he considers that time can be defined independently of motion."[105] This is followed by a passage from Aristotle: "Since time is a measure of motion, it would also be, as an attribute, a measure of rest, for every state of rest is in time."[106] The quotation, however, does not support the contention that Aristotle held that it is possible to define time apart from motion. All it says is that time is the measure of both motion and rest, which does not mean that time itself was not earlier metricized by the revolution of the heavens. Naturally, it does not follow from this that substantial temporal concepts are totally alien to Aristotle. We must remember, for example, his notion of the First Mover (form of forms), which is itself motionless but determines the dynamics of the world and "causes motion for infinite time";[107] this is the basis of absolute time. We can speak of a primary temporal substance that is uniformly distributed throughout the universe and that, after being metricized by the revolution of the heavens or other periodical natural processes, serves as the measure of motion and rest. Such a time figures in the relational conception.

Aristotle wrote:

We also know the time when we limit a motion by specifying in it a prior and a posterior as its limits; and it is then we say that time has elapsed, that is, when we perceive the prior and the posterior in motion. . . . When there are two moments . . . the one prior and the other posterior, it is then that we also say that this is time. . . . For time is just this: The number of a motion with respect to the prior and the posterior. So time is not a motion, but a motion has time *qua* the number of it.[108]

This aspect of time can be summarized by saying that to be in time is to be measured by time.

In works on the history of philosophy and the philosophy of natural science[109] Aristotle's spatiotemporal notions are widely regarded as relational, and as such they are contrasted with those of Democritus, whose void is considered the basis of the substantial concept of space. Aristotle's space as a system of relations between material objects can of course be contrasted with Democritus's void as the receptacle of material objects, but to say at the same time that Aristotle represents the relational spatial concept is to state a mere half-truth; as in the Stagirite's system, substantial space occupies an important position.

The relational concept of space in Aristotle's system has been fairly thoroughly treated in the philosophical literature and can be described as follows:

The category of space concretizes relations, specifying their character and content. Space for Aristotle is a result of object relations in the material world. He understands space to be an objective category, a property of natural things.[110]

This space is relational, but we must not lose sight of the fact that relations between material objects can also be "concretized" in Democritus's system. I have pointed out that such space, as a world structure or system of object relations, was present in most ancient Greek philosophical doctrines and characterized the world as Cosmos. In fact, both Pythagoreanism and atomism adopted just such a structural-numerical approach to the universe, so that relational space by no means contradicts the atomistic doctrine. The contradiction between Democritus and Aristotle lies elsewhere, in their approach to the void. Being an outspoken (though inconsistent) antiatomist, Aristotle objected to atoms (because nature does not take leaps), *amera* (because they conflict with continuous mathematics), and the void (because nature abhors a vacuum). Strongly influenced by such "antivoid" sentiments, Aristotle developed his famous conception of space as *topos*, that is, space as place.

What, then, is place? As Aristotle himself puts it:

That a place exists seems clear from the replacement of one thing by another, for where water is at one time, at a later time air will be there after the water is gone out as from a vessel; and since it is the same place that is occupied by different bodies, that place is then thought to be distinct from all the bodies which come to be in it and replace each other. For *that* in which air is now, in *that* there was water earlier; so, clearly the place of space into which these come and out of which they went would be distinct from each of them.[111]

Nothing can exist without place, but place can exist without anything, "for a place does not perish if the things in it are destroyed."[112] Place exists together with bodies, and all bodies are located in a place. Universal space is the aggregate of all individual places. Aristotle's thoughts on *topos* are clear, but one point remains vague: How does the *topos* conception of space differ from the Democritian notion of *kenon*, that is, from the void? At first glance there does not seem to be any difference at all. Both notions, at any rate, are modifications of the concept of space as receptacle: The void is "a box without sides," and *topos* is a vessel. Aristotle's *topos* is analogous to Democritus's void as receptacle, except that *topos* is filled not with atomic matter but with matter that is continual, continuous, voidless. Aristotle himself wrote that "those who say that a void exists include a place in their statement, since a void would be a place deprived of a body."[113] Thus refuting empty space does not imply a rejection of space as receptacle.

If, however, Aristotle's *topos* and Democritus's *kenon* are two manifestations of the same substantial conception of space, this does not at all mean that the difference between them is merely external and reducible to the "filling" contained in identical vessels. There is a considerable difference between *topos* and *kenon*. First, in contrast to the infinite void, Aristotle's space is finite and limited, as the sphere of the fixed stars spatially closes the universe. Second, if *kenon* is a passive substantial principle and merely a condition for motion, *topos* is active and possesses a specific power. This is an important distinction underlying Aristotle's dynamics, on which was erected the well-ordered if erroneous geocentric Aristotelian-Ptolemaic cosmological model that in many respects determined the development of philosophy and natural science up to the sixteenth century. Aristotle describes the "actuating" properties of space as follows:

The locomotions of physical bodies and of simple bodies (e.g. of fire, earth, and the like) make it clear not only that a place is some thing, but also that it has some power. For each of those bodies, if not prevented, travels to its own place, some of them up and others down; and these (up and down and the rest of the six directions) are parts and species of place. Now such directions (up, down, right, left, etc.) do not exist only relative to us; for to us a thing is not always the same in direction but changes according as we change our position, whichever way we may happen to turn, and so the same thing often is now to the right, now to the left, now up, now down, now ahead, now behind. By nature, on the other hand, each of these is distinct and exists apart from the others; for the up-direction is not any chance direction but where fire or a light object travels, and likewise the down-direction is not any chance direction but where heavy or earthly bodies are carried, as if these directions differed not only in position but also in power.[114]

Thus, if the void of Democritus is substantial, passive, and homogeneous, Aristotle's *topos* is finite, substantial, active, and heterogeneous. The void has no distinguished parts or directions,[115] making it difficult to speak of absolute motion, whereas in the *topos* the parts of space are distinguished, each object has its natural place, and it is in moving toward that place that the object participates in absolute motion. True, B. G. Kuznetsov maintains that "Aristotle has no notion of heterogeneous space as such; the inequality of points is due to the fact that space is filled with matter."[116] Such an assertion, however, does not fit the facts because Aristotle clearly indicates that each part of space is determined separately and possesses a certain power.

Aristotle's cosmological model operated in a finite and heterogeneous space with a distinguishable center that coincided with the center of the earth. His Cosmos is clearly divided into two levels—the earthly (sublunary) and the celestial—which did much to win him the later approval of Christian theologians. These two spheres contain completely different objects participating in completely different motions. The sublunary world consists of four elements—earth, water, air, and fire—which derive from the early pre Socratics. On their way to their natural places, these elements participate either in rectilinear natural motions (heavy bodies, for example, tend toward the center of the earth), or in compelled motions that continue so long as a moving force is exerted on them. As for the superlunary world, Aristotle thought of it as consisting of ethereal bodies in infinite, perfect, circular, natural motion. This level extends from the sphere of the moon to that of the fixed stars bounding the Cosmos. Beyond this border there is neither matter nor void.

Some of Aristotle's spatial and temporal conceptions were subsequently criticized or refined, but in general they remained unaltered for almost two thousand years and served as the basis of the Christian and medieval Cosmos. His notions mark the beginning of mature spatial and temporal metaphysics. Aristotle worked out the necessary categorical and logical apparatus of scientific inquiry and developed a fruitful concept of space and time that organically combined the static, dynamic, and cyclic levels in the structure of time and the substantial and relational levels in the structure of space. On the basis of this conception a corresponding dynamics of the world of appearances developed in which the sun revolves around the earth, bodies move so long as force is applied to them, and so on. Viewed from the vantage point of the twentieth century, these theses may seem naive. It may even be said that Aristotle's dynamics of our world is in error, but the point is that he developed a correct dynamics of his own world. The inexhaustible real world is reflected in exhaustible abstractions, and the theoretical world we will obtain depends on which abstractions we choose to include in the foundation of our theoretical constructs. The theoretical world of Aristotle rests, if you will, on superficial theoretical abstractions erected on a superficial image of the inexhaustible world. Our world image is more complex and more adequate, but let us not forget that the universe is inexhaustible, and that from its viewpoint the difference between Aristotle and us is not so great, as we live in the same world. A mere two thousand years and a few impressive revolutions in philosophy and physics are all that divide us from Aristotle. Some of these upheavals bear directly on the development and revision of spatial and temporal notions, and it is to them I turn now.

The Evolution of Conceptions of Space and Time in Medieval and Modern Philosophy

From Aristotle on, we can somewhat arbitrarily distinguish three tendencies in the development of spatial and temporal notions that do not converge until the system of Newton. What are these tendencies?

First, there is the philosophical course of development, in which Aristotle's *topos* and order were transformed through a series of intermediate steps into Newton's absolute and relative space and in which the Aristotelian primary and particular times, or eternity and time, were developed in Peripatetic metaphysics, eventually becoming the absolute and relative time of the Newtonian doctrine. If early Christian theology assimilated from ancient Greek philosophy only certain Platonic or Neoplatonic and Peripatetic notions (blunting them in the process), later Western philosophy gradually discovered the magnificent conceptual panorama of antiquity, and spatial and temporal conceptions began to draw on the entire intellectual heritage of ancient Greece. This process came to relative completion in Newton's natural philosophy, which synthesized the spatiotemporal notions of those two prominent antagonists, Aristotle and Democritus.

The second tendency has its source in Euclidean geometry, which is one of the most eminent achievements of ancient mathematics and the basis of the geometric doctrine of space. Euclid's *Elements* is an axiomatic system that was considered to be the superior approach to the organization and development of theoretical knowledge. Stressing deduction rather than observation, for centuries Euclid exerted considerable influence on the nature of philosophical and scientific inquiry. But the significance of Euclidean geometry does not end here. Combined with the temporal parameter, it served as the basis of classical kinematics, and when it was enriched by the physical concepts of mass, force, etc., it resulted in Newtonian dynamics. It was this that entitled Einstein to speak of Euclidean geometry as the oldest branch of physics. But even this is not all, for Euclid's geometry is also a certain world image. Here I can fully agree with Karl Popper and Imre Lakatos when they describe the geometry as a cosmological theory.[117] Euclid's world differs from Aristotle's; it is Platonic-Pythagorean in nature, containing the notion of homogeneous and infinite space.

The third path is the evolution of spatial and temporal concepts in dynamics. Thus in Aristotle we find the first physical dynamics possessing specific laws and principles. Intrinsically linked with this dynamics were a corresponding world image, the conception of natural places and motions, the different status of rectilinear and circular motion, their relation to different levels of the world, and so on. It is possible to speak of absolute distance between two events in space in Aristotle's dynamics, even if the temporal difference between them is not equal to zero. This accounts for the specific spatiotemporal structure of Aristotle's system, which was later reexamined and not included in the dynamics of Galileo, Descartes, or Newton. These names mark (but do not exhaust) the key points in the history of dynamics, and the spatiotemporal structures of the Aristotelian, Galilean, Cartesian, and Newtonian dynamics are amenable to reconstruction.

It must be realized that in the actual evolutionary process the above three courses of development were intertwined in a complex dialectical union. A logicohistorical study, however, cannot embrace all aspects at once but is forced to operate with certain philosophical mathematical, dynamic, and empirical reconstructions. In this section I am mainly concerned with the philosophical evolution of spatial and temporal conceptions; in the next chapter I consider the other two paths.

Post-Aristotelian philosophy contributed nothing of significance to the history of spatial and temporal concepts but basically worked with either Platonic or Peripatetic notions; the early atomists, however, had not been forgotten. At any rate, in the second and third centuries B.C. the prominent skeptic Sextus Empiricus turned to the ideas of Epicurus in his own analysis of space:

> Of the intangible nature one part is named "void" (κενὸν), another "place" (τόπος), another "room" (χώρα), the names being varied according to the different applications, since the same nature is termed "void" when destitute of any body, and becomes "room" when bodies pass through it.[118]

Thus the writers of antiquity realized that the distinction between void and place was arbitrary. It is interesting to note that a new dynamic ingredient appears here in the special mention of a space (χώρα), associated with the passage of moving bodies through the void.

As for post-Aristotelian temporal conceptions, the various schools (Skeptics, Stoics, Peripatetics and Neoplatonics) focused mainly on analyzing the interrelations between eternity and time as these were grounded in different levels of the world. For example, the Neoplatonic Plotinus (third century B.C.) gave a careful critical analysis of "false" temporal doctrines that was aimed principally at the Peripatetics.[119] He endeavored to disprove that time is identical with motion, a number or measure of motion, or a sequence accompanying motion. He tried to shift the metrication of time from the external world, where it is based on the rotational movements of the celestial bodies or other periodic motions, to the inner world of mankind. Time is "the life of the Soul in movement as it passes from one stage of act or experience to another."[120]

As Yu. B. Molchanov observes, Plotinus clearly follows the Platonic tradition of discriminating between the static and dynamic conceptions of time, contrasting motionless, timeless eternity with the rationally cognizable world of ideas, and transitory, flowing time with the "lower" world of things surrounding mankind.[121] Here we must merely bear in mind that Plotinus's eternity is immobile only when it is viewed "from below," from the world of time; he also posits another level in the world hierarchy (pure *eidos*), which is situated above eternity and from which, "from above," eternity does not appear motionless but is instead alive and in a continuous process of becoming.[122] As for the relationship between eternity and time, time is another mode of existence, or moving image, of eternity, yet it is located not so much in the world of humans as in their souls.

Here we encounter an important feature of post-Aristotelian philosophy: Mankind's interest in nature is slowly but surely supplanted by a concern with human nature, with subjective and other worldly (ideal and sacral) objects and processes. This Neoplatonic tendency was quite in keeping with Christian ideology, which entered the historical arena as a combination of Old Testament notions rooted in the mythological outlook of Oriental cultures, such as Babylon and ancient Egypt, and Neoplatonic and Peripatetic conceptions inherited from Greece. Eastern mythology exerted a sensible influence on the development of both Judaic religion and early Greek philosophy. Many Greek philosophers traveled widely through a number of Oriental lands and came into contact with the wisdom of the Median Magi, Chaldaean cosmology, Phoenician mathematics, and the cults of Egypt. These two intellectual constructs later developed relatively independently of each other, but after Alexander the Great's conquest of the Orient and Judea, Judaism became Hellenized, and this determined to a great extent the Christianity that followed. Ecclesiastes and its refrain "vanity of vanities, all is vanity and vexation of the spirit" merged organically with the unanimous choir of such post-Aristotelian Skeptics as Pyrrhon, Arcesilaus, Carneades, Agrippa, and Sextus Empiricus. Christian theology, which for almost a millennium was the only center of intellectual activity in the West, was born of the union of Athens and Jerusalem.

Church doctrine was the starting point and basis of all thought. Jurisprudence, natural science, philosophy—the entire content of these disciplines was formed in accordance with the teachings of the Church.[123]

As for the many philosophical doctrines of ancient Greece that were antagonistic toward the Christian world view, these were inherited by the Arabic East. It was Oriental philosophy that accepted the teachings of the early Greeks, commented on them extensively, elaborated many currents and conceptions, and then mediated them to medieval Europe. Even the real Aristotle came to Europe through Cordoba in the works of Averroës (the Latinized name of Ibn Rushd).

Averroës and the Averroists, however, belonged to the twelfth century; what of mankind's intellectual concerns in the preceding ages? Although Bertrand Russell observes that "the dark ages intervened between ancient and modern European civilization,"[124] the development of spatial and temporal conceptions during this period exerted considerable influence on modern philosophy and natural science. What was the prevalent medieval world image? What does the Christian Cosmos look like?

Christian theology adopted Aristotle's geocentric world as improved by the Greek mathematician and astronomer Ptolemy (second century) Aristotle's model was axiological in nature, and it was Ptolemy who in his famous *Algamest* developed a full-blown kinematic system. It now became possible to calculate the positions of the planets at any given moment. This advance involved abandoning Aristotle's concentric circles, as a complex system of epicycles and deferents had to be devised to explain obviously nonuniform and noncircular movements within a system that was presumed to be uniform and circular. The geocentric system of Aristotle and Ptolemy represented a synthesis of the old physical and cosmological conceptions of Aristotle and the new astronomical and kinematic notions of Ptolemy. It provided astronomers and astrologists with an effective means of calculation. Not only did it include a huge body of known astronomical facts but also the ease with which it adjusted to new data made it difficult to falsify the system empirically.

Theology was of course impressed by this power of calculation, but the most attractive feature of the system was its geocentricity, which seemed to offer "scientific" confirmation of Christian anthrocentrism. As mentioned, however, the system contained a variety of components; if Ptolemy's astronomy was more or less neutral ideologically, the antireligious bias of Aristotle's physics and cosmology made it vulnerable to theological revision. Spatial and temporal conceptions were especially affected by this adjustment. A detailed analysis of the spatiotemporal notions of theology might deflect us into the utterly unbelievable mystical and absurd clashes that occurred within the religious world view, and because these are of no interest to the present study I attempt to reconstruct only the main lines along which such notions developed in Christianity. These tendencies did not perish in the recesses of religious doctrine but led to the philosophical and scientific conceptions of later centuries.

The basic spatiotemporal features of the Aristotelian-Ptolemaic Cosmos can be summarized as follows: (1) Space is finite; (2) empty space does not exist; (3) time is infinite; and (4) space is divided into two levels—the superlunary and the sublunary—which obey different laws, have different structures, and do not interpenetrate.

All these theses are scientific and have been substantiated both logically and physically by Aristotle. Any scientific proposition can seem improper to the unsophisticated religious sensibility (therein the antiscientific bias of religion), as it easily encroaches on divine omnipotence. The eternity of the world postulated by Aristotle, for example, conflicted with the Christian dogma of the Creation. To make it agree with biblical creationism, time was declared to be finite. This can be regarded as introducing a Platonic element into the Aristotelian model. Quite in keeping with the Platonic tradition, early Christianity established an absolute opposition between the static celestial God (eternity) and the dynamic earthly world (time), which to some extent corresponded to Aristotle's division of the world into two levels. Already in the patristic writings, however, we find the idea of the divine trinity (Saint Athanasius at the Council of Nicaea in 325), which through the Son and the Holy Spirit included God in the sphere of the earthly world.[125]

Further, it seemed to Christian theologians that the spatial finiteness of the universe somehow detracted from God's omnipotence, for why should the Almighty create something finite and insignificant if He had it in His power to create something infinite? Interesting in this regard is the pantheism of Nicholas of Cusa, who revolutionized medieval cosmology when he proposed the boundless infinity of the Cosmos.[126] In this view the universe is theocentric rather than geocentric. It is not a geometric centration, however, because infinity has no center—the center is at any point of infinite space, everywhere because God is everywhere. In this sense E. Gilson is correct in maintaining that the Christian world view is closer to the doctrine that posits an infinite universe and plurality of worlds than it is to the ancient Greek model of a finite and single Cosmos.[127] Failing to grasp this affinity immediately, theology cruelly persecuted the advocates of plurality. The model of a single finite Cosmos also contained the notion of infinite space, but of this more shortly.

Revision of the Aristotelian Cosmos gained momentum as Christian theology became acquainted with the real Aristotle who supplanted the diluted Peripatetic fragments scattered through etymologies such as that by Saint Isidore of Seville. The genuine teachings of Aristotle were not to the taste of religious orthodoxy and were persecuted. Thus in 1210 Aristotle's *Metaphysics* and *Physics* were officially banned at the University of Paris. Another question concerned the empty space that Aristotle had so vigorously refuted. "Surely God is capable of creating a void," the theologians concluded in amazement and decreed that empty space existed.[128]

In reality, of course, these conceptual changes were not so definite, consistent, or unanimous, and they often encountered stiff resistance. Christianity contained a great many heresies and different currents that espoused diametrically opposed views (realism and nominalism, for example) and often ended their polemic at the stake. The notion of empty space did not meet with much opposition, however, because it was easily incorporated into the canonized model of the universe. The modified cosmological model of, for example, Thomas Bradwardine accommodates the void as that which surrounds the world. The concept of "nothingness," that is, the void, occupied a prominent position in theology and the patristic writings. After all, in contrast to the early Greek world view, which considered that something could not be produced out of nothing, the Christian attitude is founded on just the opposite dogma that the world was created from *nothing*. In a certain sense this tenet can be taken to indicate the primacy of empty space in the Christian world image, and it was natural that the void should be included in the Christian universe even though it contradicted Aristotle's doctrine. But in His creation God could not use up *all of nothing* (if indeed it can be used up)—the created world exists in infinite nothingness, in empty space. There are no bodies in the void, but it does contain God. Assuming infinite space in the model of the finite universe entailed assuming the existence of empty space. A number of medieval thinkers, such as Henry of Ghent, took a dim view of the notion when the Aristotelians pointed out that empty space implies infinite velocities, but on the whole theology was not discouraged by such observations. Indeed, religion needed infinite speed—how else could Saint Birgitta, for example, travel from Ireland to Italy in the twinkling of an eye? Moreover, God is not only omnipotent but also omnipresent, which is quite in keeping with the conception of divine instantaneous action-at-a-distance. This is one path leading to Newton's absolute space (the void plus the instantaneous action-at-a-distance of gravitation, that is, infinite speed), which is the *sensorium Dei*.

At first the conception of empty space wavered on the periphery of the Christian universe, but its significance grew until it acquired fundamental status by the end of the Middle Ages. Alexandre Koyré notes in this connection:

> The theological notion of divine infinity together with the notion of geometric spatial infinity gave rise to the paradoxical conception that imaginary, empty space was real—a genuine, realized nothingness in which, three centuries later, the celestial spheres holding together the beautiful Cosmos of Aristotle and the Middle Ages disintegrated and disappeared. During these three centuries the world, which had ceased to be a Cosmos, appeared to man as situated in nothingness and permeated by nothingness.[129]

Koyré has ably described the decline of the geocentric universe and the greater role allotted to infinite empty space toward the end of the Middle Ages and during the Renaissance, but objection can be made to his statement about the "beauty of the Aristotelian and medieval universes." The cosmological notions of Aristotle and the Middle Ages differed on many points, and I have examined some of those differences. But these universes also differ qualitatively with respect to their beauty. Directing special attention to this question, A. Ya. Gurevich emphasizes that, if the Cosmos of antiquity really is based on the beauty, order, and dignity of nature, the Christian Cosmos is no longer beautiful but sinful and subject to divine judgment.[130]

The notion of the Cosmos continued to function in the sense of "order." Pseudo-Dionysius, for example, maintained that the world was hierarchically ordered, a notion that enjoyed great popularity in the Middle Ages. What is meant here, however, is the mystical world and the sacred hierarchy, so that, according to St. Augustine, it is useless to seek the truth in the external world, because it is to be found in the soul and is mankind's path to God. This thesis is highly consistent with the general Neoplatonic tendency to look for the metrics of the world in the soul rather than in the world and its movements. One typical feature of the theological conception of time is connected with this point.

I have spoken of Plotinus, who rejected the metrication of time by external natural processes and transferred it instead to the inner world of mankind. These ideas were fruitfully developed by Christian theology, especially in the works of St. Augustine. Regarding the relationship of eternity and time in the spirit of Neoplatonism, St. Augustine went on to analyze the well-known Peripatetic doctrine that held that, because the past no longer exists and the future does not yet exist, all that does exist is the nondurational, instantaneous "now" of the present. He concludes that this nonexistence of the past and future obtains only from the viewpoint of the present to which we are bound. In this sense we can understand his assertion that the past and the future also exist, though in a manner that is incomprehensible to us.[131] Of the present St. Augustine observes:

As for the present, should it always be present and never pass into times past, verily it should not be time but eternity. If then time present, to be time, only comes into existence because it passeth into time past, how can we say that also to be, whose cause of being is, that it shall not be; that we cannot, forsooth, affirm that time is, but only because it is tending not to be?[132]

A characteristic change that has occurred here relative to early Greek temporal conceptions is that time has reversed its course. Notions of time in mythology and natural philosophy were connected with the image of a river flowing out of the past through the present into the future. This metaphor obviously has to do with the orientation of myth toward the past, which is the point of reference in the flow of time. Religion, on the other hand, is turned toward a coming salvation and escape from time, so that the Christian river flows out of the future. It is in this sense that we must understand St. Augustine's thesis that the present is exhausted as it moves continuously from the future to the past. Actually, here the image of the river is no longer applicable, and we must instead use that of the hourglass. God created time (He is the "creator and constructor" of time, as St. Augustine puts it) and, after measuring out a suitable quantity of it, He "poured" it into the upper chamber of the glass. This is the future, which trickles through the opening of "the present" into the lower chamber of the past. The contours of this model resemble the light cone in the theory of relativity, but in modern physics the flow or "arrow" of time has once again changed direction and returned to the ideal of antiquity—from the past toward the future; the cones of the past and future are not limited, because contemporary physics knows no unique events such as creation or salvation. Even if the Big Bang theory assumes a certain analog of "creation," it does not foresee any "salvation."

In both myth and religion time flows from the same direction toward which humans and the world aspire. In mythology emphasis is on the past, and so time flows out of it, but because mythological time is cyclic, the world is transported into an idealized past. (Incidentally, the main river of the mythological world—Okean—is circular and closed, girding the earth.) In religion the stress is on the future. Time flows from there, but because Christian time is linear, coming salvation is connected with exhaustion of the future.[133]

St. Augustine went on to develop the interesting notion of the present past and the present future, which he felt should be used instead of the traditional divisions of past, present, and future. Indeed, if the past and future exist, it is as the present. As Molchanov observes, "where they [that is, the past and future—M. A.] exist, they constitute the present."[134] Here an attempt is being made to combine the views of time from within time and from eternity. Located in time, we divide it into past, present, and future, assigning the status of existence only to the present. If the past and the future exist, they can be perceived only from eternity, but eternity lacks temporal divisions, and all segments of time belong to the present. Thus the present past and the present future are, so to speak, segments in the past and the future taken simultaneously with the present.

The reader has doubtless already noted that all these divisions are rather clumsy. They seem to be necessary to understand time, yet they become entangled in circular logic. This unnaturalness, however, is present only in the Aristotelian view, in which the temporal order is determined by the revolution of the fixed stars. In such an approach time is mechanical and mathematicized, but it has lost many of its specific features. The concept of time was greatly enriched in the subsequent history of philosophy. In the Christian conception, for example, metrication is transferred to the soul, where the past, present, and future can combine and coexist by virtue of the inherent human capacities for remembering, hoping, etc. This in turn determines St. Augustine's notion of the present past and the present future: "The present time of past things is our memory (*memoria*); the present time of present things is our sight (*intuitus*); the present time of future things our expecta-

98   *Chapter 2*

tion (*expectatio*)."[135] St. Augustine concludes that time is measured in mankind and that what is subject to measurement is not a sequence of objects but the impressions they make on the soul. Let us not forget that, according to Christian conceptions, the soul is the haven of truth and that the path to God lies through it. Thus locating time in the spiritual sphere does not diminish its importance or objectivity but indicates on the contrary that time is linked with the truth of the world through the soul of mankind.

St. Augustine, of course, could not have confidence in the order of nature and the sinful world because he knew that God creates in defiance of natural laws, that people and prophets interfere with the cosmic order (by, for example, stopping the sun), that the angels can reverse the heavens, and so on. All this is described fairly clearly in the Bible and although contemporary humans do not regard the Bible as a collection of authentic events and facts, to St. Augustine, his age, and the entire medieval period this was where the truest facts were assembled. One wonders whether modern physicists would be inclined to measure time according to the movement of the sun if they knew for certain that their colleagues were interfering with it by stopping and reversing its course. In everyday life, of course, the processes of nature provide good standards for measuring time (although this concerned hardly anyone in that distant age), but St. Augustine's interest was in the essence of time, and he arrived at the unequivocal answer that the flow of time is a psychological rather than a physical reality.

We are accustomed to thinking of subjective time as extremely relative and plural, with as many such times as there are individuals. But this is not St. Augustine's meaning—time in the soul is not subjective but absolute. As for relative time, it is determined by the periodic movements of natural objects. Time in the soul is absolute because the soul contains the truth of the world and participates in the Absolute. Thus, if God, the Absolute, is in eternity, then the soul participating in God contains or at least metricizes absolute time. Later in the history of philosophy the status of absolute space was established as the *sensorium Dei*. In this sense the Augustinian metrication of absolute time in the soul was the first attempt to determine how absolute time participated in God. Absolute time is always associated with the most perfect world objects or processes. In Aristotle's Cosmos the celestial sphere was such an object, whereas in the Christian universe it was the human soul. Absolute time subsequently becomes more intimately connected with God and is in fact likened to eternity, which acquires an inscrutable, utterly independent, uniform flow that is best described as a continuous, monotonically ascending mathematical function; this is in fact how it operates in classical mechanics. Any external metrications of this absolute time by the movements of celestial objects or the soul will yield relative time.[136] Such an understanding of time, however, was not typical of medieval philosophy, which simply attempted different approaches to the classical conception of absolute and relative time in comparative studies of "timeless being," "eternity," and "time."

"Timeless being," which Thomas Aquinas called the "measure of presence," is the immobile background relative to which time "flows." Also, Aquinas considered that the difference between flowing time and motionless "timelessness" was accidental rather than substantial.[137] In this view the static and dynamic conceptions of time in medieval philosophy are united. "Eternity" also begins to acquire a temporal character—but as the time of immutable essences. Molchanov maintains that "eternity" thus understood by medieval Scholasticism is basically identical with the "absolute time" of classical mechanics.[138] Indeed, the "eternity" of the Scholastics and Newton's "absolute time" do have something in common, but of course they also differ considerably, because they are the products of different ages and modes of thought and are incorporated in distinct conceptual and ontological schemes and dynamic systems. The mechanism by which the "eternity" of the Scholastics is transformed into absolute time is therefore of great interest. To consider this question I first turn to the spatiotemporal conceptions of the Renaissance.

Although some of Aristotle's cosmological and physical notions were revised in the Middle Ages, the geocentric Aristotelian-Ptolemaic model of the universe was sacredly canonized, and Scholasticism as a whole was thoroughly Peripatetic. The main emphasis in the anti-Scholasticism of Renaissance thinkers, therefore, was on a comprehensive criticism of Aristotle's teachings, and his doctrine of matter, space, and time was subjected to a systematic general revision. This criticism could not but affect the Aristotelian-Ptolemaic Cosmos, which was thoroughly reviewed and rejected as an artificial construct founded on false premises. It was supplanted by the heliocentric world of Copernicus, the plural and infinite universe of Giordano Bruno, and the classical system of Galileo and Newton.

Renaissance thinkers turned directly to the intellectual heritage of ancient Greece, and of course they could not be satisfied by the diluted Scholastic version of Aristotle. It would be an oversimplification, however, to reduce the Renaissance to a revival of the wisdom of antiquity. What was revitalized was the spirit of discovery in an age of major social change and epoch-making discoveries that demanded a review of established notions of the world. Geographic and cosmographic dogmas collapsed. Think, for example, of the impact of the Copernican revolution or the discovery of America! The compilations and etymologies of the Scholastics met with a healthy methodological skepticism as observation and experiment, geometry, and kinematics came into their own. The foundations of experimental science were laid, and knowledge became power. As Engels observed:

Modern natural science . . . begins with that mighty epoch when feudalism was smashed by the Burghers. In the background of the struggle between the Burghers of the towns and the feudal nobility this epoch showed the peasant in revolt, and behind the peasant the revolutionary beginnings of the modern proletariat, already red flag in hand and with communism on its lips. It was the epoch which brought into being the great monarchies in Europe, broke the spiritual dictatorship of the Pope, evoked the revival of Greek antiquity and with it the highest artistic development of the new age, broke through the boundaries of the old world, and for the first time really discovered the world.[139]

The fertile and revolutionary beginnings made by the Renaissance thinkers encountered increasing opposition on the part of the Inquisition. To dethrone Scholastic dogmas and develop a new theory of the universe, recourse was had not only to the conceptions of antiquity but also to various anti-Aristotelian notions that had evolved within medieval philosophy, in which could be found—often as heresies—concepts of infinite space, the void, finitism, and so on.

Renaissance thinkers drastically revised the Peripatetic view of the world canonized by Scholasticism, particularly Aristotle's doctrine of space and time. The static model of space was poorly suited to the basic attitudinal tendencies of the period, which devoted considerable attention to perspective, dynamics, and the like.[140] A. Kh. Gorfunkel' has noted of this transition in spatial notions: "The absoluteness of Aristotle's 'natural' place was opposed by an absolute concept of space as the objective reality in which motion occurs."[141]

The notion of uniform space systematically developed by the Renaissance derived from early Greek atomism and the void of Democritus. In Bernardino Telesio, for example, we find the characteristic idea that, although no part of the universe is empty, this does not in itself imply a refutation of the void.[142] Aristotle could not solve this paradox and held that, if everything is filled, one can speak only of *topos* and not of empty space. Telesio subtly perceived the inner unity of these two approaches and emphasized the substantiality of the void. This was connected with the developing notion that uniform space is a necessary condition for the existence and dynamics of matter. Here the Renaissance philosophers were no longer satisfied by logicotheoretical proofs of the existence of the void but attempted instead to base its reality on experimental physical evidence. Telesio holds that, although the world may be completely full, this does not detract from the reality of the void; as is experimentally demonstrable, it can be formed by applying physical forces or actions. Telesio's experiments attracted the attention of such contemporaries as Francis Bacon, who comments on one experiment that was proposed to prove the existence of the void:

> This he tries to prove by certain experiments, especially adducing those which are everywhere cited for the contradiction and refutation of a vacuum, and as it were making extracts from them, and amplifying them so as to allow beings to be under some slight necessity of holding to that which is contiguous, but so that if they be more strongly pressed, they will admit a vacuum; as we see in water-clocks, in which if the hole through which the water runs is too small, they will want an air-hole to enable the water to descend; but if the hole be larger, even though there be no air-hole, the water, pressing with a heavier weight on the hole, flows downwards, not caring for the vacuum above.[143]

Telesio's conceptions were elaborated further by Francesco Patrizi, who clearly declares space to be primary:

> If a substance is that which exists in itself, then space is a substance in the highest degree, for it exists in itself in the highest degree.... It

requires nothing for its being, but all other essences require it for their existence. . . . It is the most primary of all essences.[144]

Bruno also developed the substantial notion of space as receptacle. If Telesio's first steps toward the idea of homogeneous space had remained within the limits of a finite and geocentric cosmos, Bruno was moving toward the notion of infinite space. Space is not primary as a genetic principle; its primacy is instead logical and substantial, because space is a necessary condition for motion and the existence of matter: "For a body does not exist except as somewhere; it cannot exist if there is no space. And this is the void."[145]

This fertile development of the substantial conception, which in Bruno attains the notion of homogeneous and infinite space, contributed to a thorough revision of the Aristotelian-Ptolemaic geocentric system, in which space had been heterogeneous and finite. This system was smashed, and a new world was created on its ruins! The Renaissance abounded in discoveries, but nothing can compare with the Copernican revolution. The upheaval it caused was so enormous that it was not immediately noticed. Copernicus's *On the Revolution of the Heavenly Spheres* was at first regarded as a convenient apparatus of calculation, a view that was also suggested by the theologian Andreas Osiander's misleading foreword to the book. Copernicus, however, did not merely recenter the cosmic system by exchanging the sun for the earth but radically altered the universe itself, transforming its spatial structure and the entire conceptual system of the old cosmology.

The geocentrism of the theological Cosmos was anthropocentric, and the universe itself was divided into various celestial, sacred, and earthly or sublunary levels. This opposition of the earth and the heavens was physical (different spaces, times, and motions), chemical (the levels consisted of different elements), and moral and ethical. The cosmic order was merely an imperfect copy of the sacred world hierarchy. Copernicus's cosmology razed this entire structure to its foundations. Relations among cosmic objects became purely physical—"the earth" and "the heavens" are subject to the same laws; the Cosmos is a unified construction to which correspond the notions of a single homogeneous space and relative motion. The next item on the agenda was the creation of a new mechanics, which was undertaken later by Galileo and Newton.

Copernicus could not fully rid himself of the idea that the universe is finite and limited. His Cosmos is bounded by an eighth sphere in which all stars are equidistant from the center. Although he thought of the world as spherical and limited, he extended its spatial boundaries so greatly that he was close to recognizing its infinity. There is the characteristic statement, for example, that "the massive bulk of the earth does indeed shrink to insignificance in comparison with the size of the heavens. . . . The universe is spherical, immense and similiar to the infinite."[146]

Copernicus apparently felt that revising the old cosmology would lead to a review of finite space and of the Cosmos itself. He was also absorbed, however, by the problem of calculation. In his central model of circular motions, epicycles, and celestial spheres, he addressed the necessary task of reckoning the entire mechanics of the new universe. Bruno continued, surmounting Copernicus's limitations and bringing his undertaking to its logical conclusion—a plurality of worlds in infinite space. Bruno wrote:

I insist upon infinite space, and nature itself has unlimited space not by virtue of its dimensions or corporeal volume, but by virtue of nature itself and its kinds of bodies; for divine superiority is incomparably better represented in innumerable individuals than in those that are numerable and finite.[147]

An interesting shift has occurred in the approach to space. In Aristotle's day it was possible to speak of a finite Cosmos bounded by the sphere of fixed stars without ever asking what lay beyond that sphere. In the Renaissance this question inevitably arose, leading Bruno to observe that a finite universe is inconceivable outside a void that extends at least beyond the limits of the finite Cosmos.[148] This, however, was merely an idea, not the core of Bruno's cosmology. That foundation was the Copernican revolution, from which Bruno drew no less revolutionary conclusions.

The new heliocentric world system created by Copernicus still contained certain Peripatetic and Scholastic elements. Bruno banished these as well from the new Cosmos, exploding the limiting sphere and opening up the universe to infinite space. He deprived the Cosmos of its center and laid the real foundation of the homogeneous infinite space within which classical physics was born.

After the accomplishments of the Renaissance thinkers, the few wretched shreds that remained of the theologians' Cosmos hardly sufficed to fuel the Inquisition's bonfire on the Campo dei Fiori in 1600. The seventeenth century began dismally—the furious Nolan was no more, but Kepler, Galileo, Descartes, Newton, and the other brilliant philosophers who succeeded him completed the rout of Scholasticism, established the laws of motion of the new universe, and laid the cornerstone of modern science.

Before proceeding to an analysis of spatiotemporal conceptions in the natural philosophy of the founders of classical mechanics, however, let us once again return to the Renaissance and consider the temporal notions of that age. Copernicus's approach to time is of indisputable interest, as his views mark a turn from Scholastic metaphysics to the real physics of time. He was especially interested in the uniform flow of time and the means by which it could be determined and measured. He realized that the natural succession of days and nights could not provide a universal and precise measure of motion, because these 24-hour periods are not identical in all respects. "Therefore," he wrote, "it was necessary to select from these ⟨days⟩ some mean and uniform day which would permit uniform motion to be measured without uncertainty."[149] Copernicus devised one of the first formulas for calculating mean astronomical time. Molchanov correctly observes that it enabled him to discover the empirical basis of the "true" mathematical time of classical mechanics.[150] As to the logicomethodological foundation of this notion, it had already been laid by medieval philosophy and developed by such Renaissance thinkers as Telesio: Time is objective and exists independently of motion or changes in material objects.

Many philosophers and naturalists subscribed to such a notion of absolute time. Among the precursors of Newton (who based his famous *Principia* on absolute time and space) is Pierre Gassendi, who articulated the then prevalent conception as follows:

I, at least, know one single time, which, of course (I do not deny it), may be called or considered abstract, since it does not depend on things; for whether things exist or not, whether they are moving or in a state of rest, it always flows uniformly, subject to no changes whatever.[151]

Thinkers of an Aristotelian persuasion had looked to external nature to find temporal metricizers, such as the revolution of the sphere of fixed stars, but in an infinite universe that approach becomes meaningless. As Bruno demonstrated, in such a Cosmos all scales and metrications of time are relative and depend on the observer because there do not and cannot exist any limiting spheres capable of providing the events of the world with a universal and absolute rhythm.[152]

It is important to note one significant innovation in the way in which the Renaissance formulated the problem of time. As the heliocentric world system evolved, the division into earthly and celestial levels was abolished and the universe was unified. Let us recall, however, that this distinction between "earth" and "the heavens" was at the same time a division into time and eternity. Thus the Copernican revolution created the necessary prerequisites for a combined analysis of time and eternity as equally valid aspects of a unified world. And although modern theologians continue to maintain that eternity is not simply infinite time but exists outside of time and is therefore intrinsic to God alone,[153] the contemporary scientific view of the relationship between time and eternity is in line with the Copernican union of "earth" and "the heavens." "Infinity" is a metric property of time that characterizes duration; eternity is a topological property that describes the "organization" or structure of time "on the whole" or "as a whole."[154]

The new heliocentric world system was born in the sixteenth century. The genius of Copernicus was needed for its creation, but understanding and developing it further required at least the no less talented and daring Giordano Bruno. Such men were few, however. Some were intimidated by the Inquisition, whereas others simply could not grasp the new system because it was too unusual and too obviously contradicted immediate sensations (anyone can observe the rising and setting of the sun), the authority of the Bible (Joshua stopped the sun, not the earth), and a great many theoretical and physical theses. Thus Ptolemy argued that the motion of the earth would disrupt natural processes—clouds and birds would be carried toward the west, falling bodies would deviate in that direction, and so on. All these problems were inherited by the seventeenth century, when a constellation of brilliant philosophers, physicists, and mathematicians—often all three in one and the same person—set about substantiating, proving, and elaborating Copernicus's theories.

The first step in this direction was taken by Johannes Kepler, who in *Astronomia nova, seu physica coelestis* (1609) and *Harmonices mundi* (1613) set forth the three laws of planetary motion that made him famous: (1) The planets describe elliptic orbits, of which the sun occupies one focus; (2) the line joining a planet to the sun sweeps out equal areas in equal times, and the area of the orbital sector described by the radial vector of the planet varies proportional to time; (3) the square of the period of revolution of a planet is proportional to the cube of its average distance from the sun.

Thus, contrary to the Scholastic dogma of perfect circular orbits, Kepler introduced elliptical orbits, an innovation of great methodological significance that greatly furthered the intellectual emancipation of Europe. Kepler refuted the deification of one element of geometry—the circle or sphere—but went on to deify geometry itself, the mathematical and physical doctrine of space. He not only thought that "the traces of geometry are stamped on the world as if geometry was the prototype of the world," but also considered that "geometry is God Himself."[155] Accordingly, he developed the notion that human perception is instinctively bound to Euclidean geometry. Here I should distinguish the space of the instincts (which may very well be Euclidean) from the instinctual adherence of perception to Euclidean geometry, which (as we saw in chapter 1) does not correspond to either modern psychological concepts or (as we will see in chapter 3) the conceptions of contemporary physics.

Two spatial conceptions originating in Kepler were developed further by later thinkers. On the one hand, Kepler's deification of geometry appears in the intimate association of space and God in Newton's natural philosophy; on the other hand, his archetypal spatial instinct was transformed by Kant into the notion of space as an a priori form of sense intuition.

Kepler's dynamics was to an extent influenced by the Aristotelian doctrine and held that a moving object would stop unless it was kept in motion by a force from without. He did not go as far as the idea of inertia. This notion was first advanced by Galileo, whose *Dialogue Concerning the Two Great Systems of the World* contributed enormously to substantiating the heliocentric system.

Both empirically and theoretically, Galileo showed that the Aristotelian universe was untenable. With the telescope he invented, he made remarkable discoveries that vividly demonstrated how distant the Scholastic world was from reality. There are mountains and craters on the moon, the planets have satellites, and the Milky Way is a giant conglomeration of stars. A Cosmos limited by the solar system was out of the question. Galileo went on to tackle problems of mechanics, disproving Aristotle's idea that a body in the void would move with infinite velocity. (In Aristotelian dynamics, speed is directly proportional to exerted force and inversely proportional to density of environment.) Galileo's quite different concept held that bodies fall in a vacuum at the same finite speed.[156] This annulled one of Aristotle's most weighty arguments against the reality and logical possibility of empty space. In his later elaboration of this conception Galileo analyzed the motion of bodies along an inclined plane and formulated the principle of inertia: "[The motion of] a movable object projected on a horizontal plane with all impediments removed . . . will be equal and perpetual on that plane, if the plane be extended to infinity."[157] Here it might seem as though Galileo is speaking of rectilinear inertial motion (incidentally, that is Feynman's opinion,[158] but in fact he is not. He was considering circular inertial motion, and in the passage quoted he is referring to the surface of the earth.

I have violated chronology somewhat by beginning to treat Galileo's conception with his refutation of the Aristotelian objections to empty space, as I think this was a decisive point in the development of his spatiotemporal notions. As B. G. Kuznetsov correctly notes, the concept of empty space provided the prerequisites for seventeeth-century dynamics.[159] As for Galileo's inertial circular movements, the entire celestial mechanics of the *Dialogue* is based on them, although the movement of celestial bodies could not really be explained until Descartes's concept of rectilinear inertia.

One gets the impression that the great triad, Galileo, Descartes, and Newton, was discussing different combinations of spatial concepts and inertia. As I noted, Galileo recognized empty space and circular inertial motion, and Descartes went on to the idea of rectilinear inertial motion but rejected empty space. It was not until Newton's natural philosophy that these two indispensable components of classical mechanics were united. And although Newton rejected or "removed" the Cartesian doctrine of space, time, and motion, much of what Descartes had developed prepared the way for his own theories.

Descartes contributed a great deal to geometry and the geometric study of nature. He advocated uniting physics and geometry, regarding their union as a goal of science. He introduced the coordinate system bearing his name in which time represented only one of the spatial axes in the form of a straight line. This was convenient and stimulated the rapid growth of scientific knowledge, but it also assigned inappropriate spatial features to time. Engels fully appreciated the revolutionary novelty of analytic geometry:

The turning point in mathematics was Descartes' *variable magnitude*. With that came *motion* and hence *dialectics* in mathematics, and *at once also of necessity the differential and integral calculus.*[160]

Differential and integral calculus was developed by Newton and Leibniz. As for the Cartesian geometrization of matter, it entailed the identification of material substance with extension. Descartes thought that the true essence of bodies was expressed not by sense images but by their logicomathematical properties; this is why he regarded the geometric feature of extension as the chief attribute of matter and why he identified space with material extension. This thesis forced him to reject empty space, because otherwise he would have had to admit the existence of nonmaterial extension. The contradiction between Cartesianism and atomism that this denial appears to imply is only a seeming one,[161] for as was shown in the preceding section, the void of the atomists did not have the property of extension. Descartes, by contrast, treated empty space as a nonmaterial extension, and it is with reference to this status that he rejects its reality. He observed:

That a vacuum in the philosophical sense of the term (that is, a space in which there is absolutely no [material] substance) cannot exist is evident from the fact that the extension of space, or of internal place, does not differ from the extension of body [here again *kenon* is rejected in favor of *topos*—M. A.]. From the sole fact that a body is extended in length, breadth and depth we rightly conclude that it is a substance: because it is entirely contradictory for that which is nothing to possess extension. And the same must also be concluded about space which is said to be empty: that, since it certainly has extension, there must necessarily also be substance in it.[162]

Descartes developed further the notion of two times. As we saw, from antiquity and on through the Middle Ages and the Renaissance there existed a concept of a first, absolute time (sometimes viewed as eternity) and its empirical analog, relative time, obtained through the revolution of the celestial spheres, a succession of images in the soul, etc. Descartes treated this notion in the context of duration and time. As articulated in his *Principles of Philosophy*, time as a mode of thought corresponds to objective reality:

However, some attributes or modes are in the things themselves, while others are only in our minds. Thus, when we distinguish time from duration taken in general and say that time is the measure of motion, this is only a mode of thinking. ... In order to measure the duration of all things, we compare it with the duration of the greatest and most uniform motions from which days and years are created, and we call this duration "time": which, accordingly, adds nothing to duration taken in general except as a mode of thought.[163]

The entire temporal complex of both Descartes and a number of other seventeenth-century thinkers, such as Spinoza,[164] can be described as consisting of three components: eternity, which is intrinsic to God; duration, which is intrinsic to the material world; and time, which as a mode of thought is intrinsic to mankind. This temporal conception was developed further by Newton, who related it organically to mechanical theory. Time as viewed by Descartes, however, is not exhausted by the given thesis on duration. Descartes was a dualist, after all, and underlying many problems solved in a materialistic spirit one often finds a theological or idealistic hypostasis. This is the case with Descartes's temporal conception. The notion of discrete time advanced in the *Meditations on First Philosophy* holds that God directly intervenes with each successive discrete instant and that such interference is the cause of the entire diversity of natural objects. Descartes notes:

For all the course of my life may be divided into an infinite number of parts, none of which is in any way dependent on the other; and thus from the fact that I was in existence a short time ago it does not follow that I must be in existence now, unless some cause at this instant, so to speak, produces me anew, that is to say, conserves me. It is as a matter of fact perfectly clear and evident to all those who consider with attention the nature of time, that, in order to be conserved in each moment in which it endures, a substance has need of the same power and action as would be necessary to produce and create it anew, supposing that it did not yet exist.[165]

Thus, proceeding from general theological conceptions, Descartes regarded the discreteness of time as indicating the omnipotence of God, who at every moment, or "atom," of time performed an act of creation (or rather conservation), generating the entire substance and variety of nature out of nothing. Whitrow notes of this aspect of Cartesianism that a material body has only the property of spatial extension and no inherent capacity for endurance, so that at each successive instant God must recreate the body to prevent it from continually slipping into not-being.[166]

It would be a serious error to separate Descartes's physical and metaphysical (and sometimes simply theological) notions of space and time by assigning them different levels of being or gradating them as "earthly" or "celestial," because these ideas are intimately interrelated and operate within a unified conception. The temporal concept connected with continuous acts of creation or conservation, for example, was central to Descartes's mechanics and brought him (in contrast to Galileo) to the possibility of rectilinear inertia and laws of motion that are formally similar to Newton's. It will suffice to cite two of them.

According to the first law of nature, each thing, insofar as is in its power, always remains in the same state; and consequently, once it is moved, it continues to move.

The second law states that all movement is, of itself, along straight lines; and consequently, bodies that are moved in a circle always tend to move away from the center of the circle they are describing.[167]

Here we have an interesting phenomenon. Descartes formulated laws of motion and a version of the principle of inertia that formally coincided with Newton's, yet he based these laws on theological notions and the world to which they belonged. The first law, for example, holds because God is immutable and conserves equal amounts of substance and motion in the universe. What was needed was a natural-philosophic revision of the Cartesian concepts of space, time, motion, and matter. This was undertaken by Newton, who in the process paradoxically arrived at Descartes's laws. These, however, were incorporated in a different system of natural philosophy and were derived by a different procedure within a distinct physical mechanics. Strictly, they are different laws in a different universe.

Commenting on Descartes's and Newton's coinciding laws of motion, Richard J. Blackwell correctly observes that the meaning of a scientific law includes much more than a description of the physical state of affairs; also involved is an understanding of why the designated state of affairs obtains. Blackwell calls this factor the theoretical meaning of a scientific law.[168] Thus there can exist two versions of a law that coincide on the level of descriptive meaning but that differ considerably on the theoretical one. This is exactly what we find in the laws of Descartes and Newton.

Descartes does not consciously or systematically take the relativity of motion into consideration. His notions revolved within the conception of geometrization, and Newton's idea of mass as inertial resistance to change was completely alien to Descartes. The only inseparable property of matter that Descartes recognized was extension, and he would therefore have attempted to reduce mass as well to the spatial volume of material objects. The dynamic conception of mass was of enormous significance to Newton's system, however, and herein lies an essential difference between his and Descartes's laws of motion. Blackwell notes in this regard:

At the descriptive level [Descartes's] first two laws of nature designate the same state of affairs as Newton's first law of motion, but the theoretical differences are immense. A body perseveres in its state of motion or rest for Descartes because this is demanded by the immutability of God. The same holds true for Newton because of the body's mass. In the first case the reason is external, in the second internal, to the material world.[169]

Thus, whereas Descartes can be described as a representative of the old metaphysics who approximated the new physics and, at least on the descriptive level, even anticipated some of its fundamental propositions, Newton is the creator of a new physics that determined his natural philosophy and theory of space and time. Of course, this does not imply that he broke with theological metaphysics. On the contrary, he was a child of his age who was even actively concerned with theological problems; it was not these, however, that defined his philosophy of nature, which rests on ancient Greek atomism and the anti-

Scholastic and anti-Peripatetic ideas of his intellectual predecessors, such as Galileo, and his teachers, such as Isaac Barrow.[170]

The atomistic theory of matter as developed by Newton[171] therefore contains empty space as an indispensable component, and in this regard Newton stands in sharp contrast to Cartesianism. Furthermore, his physical dynamics operates on a physical (and not a theological!) principle of inertia and on a dynamic (not a spatially static!) concept of mass. This does not mean, however, that Newton rejected all of Descartes. In his mathematical analysis of space and time, for example, he continued along the Cartesian line. Descartes had introduced the concept of the variable. Greatly admiring the achievement, Engels remarked that it was this that made differential and integral calculus necessary. This "order" submitted by Descartes was brilliantly filled by Newton, whose physical dynamics required just such a mathematical apparatus. The differential nature of the Newtonian laws of motion distinguished them significantly from the laws and concepts of his precursors. For the first time in science it became possible to take a state of motion at a given moment in time and deduce from it the state immediately following it. Commenting on this, Einstein wrote:

The clear conception of the differential law is one of Newton's greatest intellectual achievements, [for] the differential law is the only form which completely satisfies the modern physicist's demand for causality.[172]

Newton's system distinguished two types of space and time—absolute and relative. What, exactly, are they?

Absolute, true, and mathematical time, of itself and from its own nature always flows equably without relation to anything external and by another name is called duration. . . . Absolute space, in its own nature, without relation to anything external, remains always similar and immovable.[173]

Thus absolute space is the empty receptacle of material objects. Does Newton's absolute space possess any positive properties that distinguish it from the void of the Greek atomists, which is "nonextension"? The argument associated with the nonextension of Newton's absolute space is not new and derives from his chief adversary, Leibniz. Thus, in his polemic with the Newtonian Samuel Clarke, Leibniz observed that extension is a property of the extended, whereas the void is nonextended.[174] From the viewpoint of such a conception, continuous absolute space contains only a potentiality of extension and is metrically amorphous, because a metric is introduced into it in a purely external fashion. Absolute space is in fact not extension but the abstract condition for the existence and motion of material objects. Analogously, the Democritian void, because it lacks the attribute of extension, is the necessary condition for (but not the source of) the motion of material atoms. It would be naive to suppose, however, that Newton's absolute space is fully analogous to Democritus's void. The features of absolute space organically incorporated the structure of classical mechanics derived from a different age, other forms of thought, other epistemological and methodological approaches, and a different world image. Classical mechanics is in fact based on absolute space (and time). The void, of course, is nothingness; but where the biblical deity managed to come up with enough of it to create the world, physics derives from nothing the fundamental laws of the world, for the basic laws of classical physics are based on the properties (homogeneity, isotropism) of absolute space and time. Commenting on the conservation laws, Kenneth Ford notes that they follow from the simple fact that empty space lacks features and is everywhere equally empty and equally undistinguished. "It seems in the truest sense," he says, "that we are getting something for nothing."[175]

This is what gives absolute time and space their theoretical basis; it also disarms criticisms of the notions, which though common are generally based on empiricist and positivist arguments. This entire set of problems is dealt with in chapter 3. Let me for the moment discuss what distinguishes Newton's absolute space from that of Aristotle.

Newton systematized and brought to completion Copernicus's and Galileo's destruction of the Aristotelian system of natural space on which absolute space is strung. This concept loses its clear kinematic meaning in Newton's system, where, as B. G. Kuznetsov notes, absolute space becomes meaningful only because observations can register absolute motion by the appearance of forces of inertia in acceleration.

Thus instead of absolutely immobile bodies of reference and kinematically graphic absolute space Newton introduces a local criterion: the behavior of an accelerating body at a given point differs from that of a motionless or uniformly moving body.[176]

It is this criterion that lends Newton's absolute space its novelty and distinctiveness.

Parallel with absolute space Newton also considered absolute time, which seems a most paradoxical object. Let us recall that, in Newton's view, absolute time, from itself and without regard to anything external, flows equably and is otherwise known as duration. This definition is paradoxical first of all because of the logically unsatisfying detail that the flow of time is associated with the representation of time as a process in time; second, as Whitrow correctly notes, it is difficult to accept the statement that time flows uniformly, because this seems to imply that there is something that controls the rate of flow. Moreover, if time is considered

"without relation to anything external," what meaning can be attached to saying that its rate of flow is *not* uniform? If no meaning can be attached even to the possibility of non-uniform flow, then what significance can be attached to specifically stipulating that the flow is "equable"?[177]

Here, however, it must be borne in mind that Newton referred to absolute time as not only true but also mathematical. On this point his temporal conceptions differ significantly from the view of his predecessors. Before Newton, by true time was meant sometimes simply duration, sometimes infinite time, and sometimes eternity itself understood as infinite time in its static totality but not as timelessness. This time was then metricized through periodic natural movements, in which case it was regarded as participating in Nature, or it was measured in the reminiscences and expectations of the soul, in which case it became a psychological construction or mode of thought that yielded relative time. With Newton, an intermediate level appears: Whereas earlier concepts operated with binomial oppositions, such as "God–Nature" or "God–man," Newton speaks of the trinomial construction "God–theory of Nature–Nature"—hence the intermediate character of the metrication (mathematical and theoretical). Metrication transforms time as eternity into time of the theoretical level of physics. This, then, is mathematical time, whose duration is determined by the line of Euclidean geometry and whose rhythm is defined by a natural numerical series. Time is a monotonically increasing function in the interval $(-\infty, +\infty)$.

By accepting absolute time, Newton also had to postulate absolute and universal simultaneity. The clockwork is the same throughout the universe. Only forces acting at a distance can serve as the basis of absolute synchronism. The role of such forces was assigned to gravitation, which was elevated to the rank of a universal (the universal law of gravitation).

The nonphysical character of action-at-a-distance disturbed even Newton, who attempted to develop a short-range theory of gravitation. The crux of the matter, however, is that action-at-a-distance is connected not with gravitation but with the conception of space and time. As Kuznetsov notes:

In Newton's physics forces were distributed rectilinearly over infinite distances. These Newtonian forces—the instantaneous, extratemporal interactions of bodies—were the physical framework of Newtonian space, which existed independently of time.[178]

Here there is also a certain paradox that in general inevitably accompanies absoluteness (absoluteness is internally paradoxical). Thus action-at-a-distance, which was incorporated in Newton's system to substantiate absolute space and time, repudiates the very idea of space and time. Acceptance of action-at-a-distance denies the continuity of absolute space and time, introducing into them absolute disruption.

The extratemporality of forces acting at a distance is intrinsic to their instantaneous essence. Emile Meyerson has also discussed their extraspatiality:

The hypothesis of action at a distance consists in supposing that one phenomenon is the condition of another and that nothing happens in the intermediary space. Assuredly it will be affirmed that that intermediary space is traversed by force. But the two phenomena being simultaneous, force does not cross space; it leaps over it, if we dare to express it thus.[179]

This leads Meyerson to conclude that the idea of instantaneous action-at-a-distance is destructive of the concept of space.

Action-at-a-distance can also be regarded with respect to infinite velocity, and here A. F. Losev presents an interesting analysis:

But what is this body moving with infinite speed? It means that (1) the body is everywhere at once, in all places through which it is capable of passing; (2) these places, even if there were an infinite number of them, no longer seem to exist for the moving body, since it has them all in itself; (3) consequently, the motion of a body at infinite speed is equal to rest, since there is no place in which the body could not be located and no place to which it could still move.[180]

Losev's argument clearly demonstrates the abstract identity of absolute rest and absolute velocity—zero and infinity meet and coincide. Incidentally, the theory of relativity has confirmed this interconnection—the rejection of infinite speed inevitably led to the rejection of absolute rest.

From absolute space Newton distinguished relative space, which is the duration of material objects. The Cartesian extensional conception of space operated at the empirical level of the Newtonian system. Relative space served as the measure of absolute space.[181] Relative time was introduced corresponding to relative space:

Relative, apparent, and common time is some sensible and external (whether accurate or unequable) measure of duration by the means of motion, which is commonly used instead of true time; such as an hour, a day, a month, a year.[182]

Relative space and time are the metricized analogs of absolute space and time.

Newton's conception of space and time held complete sway until the end of the nineteenth century, and it was not until the advent of electromagnetism that its limitations began to become apparent. The twentieth-century theory of relativity re-examined, or rather modified and significantly developed, Newton's notion. All this is discussed in chapter 3; for the time being, I dwell briefly on the spatiotemporal concepts of Leibniz, Kant, and Hegel, as the original ideas of these thinkers variously influenced notions developed by natural science and dialectical materialism.

Leibniz's views are of indisputable interest in their own right and not just as a critique of Newton. As we know, Leibniz developed the notion of relational time, and we are justified in asking wherein its originality lay, especially because Newton also introduces relative space and time at the empirical level. Leibniz's concepts derive in many respects from his monadology. He proposed three minimums—atoms, points, and monads—corresponding to the physical, mathematical, and metaphysical levels. Leibniz wrote:

Physical points are indivisible in appearance only: mathematical points are exact but they are nothing but modalities. It is only metaphysical points, or points of substance (constituted by forms or souls), which are both exact and real; and without them there would be nothing real, since without true unities there would be no plurality.[183]

Thus Leibniz conceives of extension not as the sum of nonextended monads but as the result of the active dynamic being of these essences. Natural science adopted this notion and transformed the theory of discrete spiritual essences into the concept of discrete centers of force on which dynamic atomism is based.

Leibniz's notion of relational space rests on the idea that

one monad, in itself and at a particular moment, can only be distinguished from another by internal qualities and activities, which can be nothing else but its *perceptions* (that is to say, the representations in the simple of the compound or of that which is outside) and its appetitions (that is to say, its tendencies to pass from one perception to another), which are the principle of change.[184]

Each monad mirrors the diversity of nature in its three-dimensional perspective, or perceptual space; "Each monad is a mirror that is alive or endowed with inner activity that is representative of the universe from its point of view and is as much regulated as the universe itself"[185] in accordance with a pre-established harmony.

Analyzing Leibniz's conceptions, Bertrand Russell concludes that his system contains two types of space:

What I, for my part, think best in his theory of monads is his two kinds of space, one subjective, in the perceptions of each monad, and one objective, consisting of the assemblage of points of view of the various monads.[186]

Aristotle had proposed natural places to counterpoise the Democritian void. When all such places were combined, however it became possible to speak of a single Place framed by the celestial sphere. This limiting Aristotelian space was the same Democritian receptacle, except that it was continually filled. At the limit Leibniz's second space also seems to tend to become an analog of Newton's absolute space. The plurality of gods, that is, Leibniz's monad, can by joining their perceptual forces recreate that which is in the power of Newton's one God, whose sensorium is absolute space. Incidentally, in one way or another Leibniz also assumed absolute space among other places in his fifth letter to Clarke, leading scholars such as Stallo and Reichenbach to accuse him of inconsistency.

Proceeding from the principle of pre-established harmony, Leibniz postulated an identical temporal order for the monads—they were of necessity synchronized because each of them mirrored the universe. As Whitrow correctly observes, "insofar as the temporal aspect of the universe is concerned, Leibniz's principle of harmony is equivalent to the postulate of universal time."[187]

Thus, although Leibniz's conception of space and time is relational, his system also at least potentially contains the notion of absolute space and universal time. As for his relational concept, it is articulated most clearly in his polemic with the Newtonian Clarke. In the third paper Leibniz writes:

I have said more than once, that I hold space to be something merely relative, as time is; time is an order of successions. For space denotes, in terms of possibility, an order of things which exist at the same time.[188]

In developing his idea, Leibniz touches on the interrelationship of space and matter:

I don't say that matter and space are the same thing. I only say, there is no space, where there is no matter; and that space in itself is not an absolute reality.[189]

Thus, according to Leibniz, space and time do not exist either in their own right as substances or as identical with matter but are forms of existence of material objects, abstract structural relations of their existence. This conception did not exert any great influence on classical mechanics, but methodologically the presence of two rival notions (the substantial and the relational) significantly affected the evolution of science and philosophy by stimulating broader and more adequate views on the nature of space and time.

It is on this level that we should regard the notions of Kant, who in his pre-Critical period advanced an important cosmological hypothesis and in the later Critical period developed an original spatiotemporal conception. Kant's cosmology was of immense significance to the development of science and its world view. For the first time the origin of the solar system was subjected to a scientific study, as opposed to a mythological or theological study. The title of his work—*Universal Natural History and the Theory of the Heavens*—speaks for itself. This study represents a change in approach to such universal problems. Kant writes:

It seems to me that we can here say with intelligent certainty and without audacity *"Give me matter, and I will construct a world out of it!"* i.e. give me matter and I will show you how a world shall arise out of it. For if we have matter existing endowed with an essential force of attraction, it is not difficult to determine those causes which may have contributed to the arrangement of the system of the world as a whole.[190]

The importance of Kant's cosmological theory cannot be overstated. As Engels notes, it was the greatest achievement in astronomy since Copernicus:

For the first time the conception that Nature had no history in time began to be shaken. . . . Kant made the first breach in this conception, which corresponded exactly to the metaphysical mode of thought, and indeed he did it in such a scientific way that most of the proofs used by him still hold good today.[191]

The entire Kantian cosmological system is based on Newton's theory, which in Kant's treatment acquired a shade of absoluteness and paved the way for his later apriorism. Analyzing this aspect of Kant's relationship to Newton, Hans Reichenbach subtly notes that Kant "regarded the physics of Newton as the ultimate stage of knowledge and idealized it into a philosophical system."[192] Kant's cosmological system accordingly contains the notion of empty space.

In subsequent works, such as *The Application in Natural Philosophy of Metaphysics Connected with Geometry*, however, Kant turns to physical monadology, which brings him to a different understanding of space. Asserting that there are two approaches to space, he points out their limitations:

Those who assume the infinite divisibility of real space have absolutely refused to recognize the existence of monads, and conversely, the defenders of the monads have for their part found it necessary to regard the properties of geometrical space as something imaginary.[193]

Kant himself endorses the following point of view:

> But since space is not a substance, but merely a kind of appearance of the external relations of substances, the possibility of dividing the relations of a single substance into two does not contradict the simplicity or, if you will, the unity of the substance.[194]

And a little later he explains:

> Since space is composed exclusively of external relations, nothing internal in a substance . . . is determined by space itself; in space we are only justified in looking for those determinations of the substance which belong to its external aspect.[195]

Kant never abandoned these notions of space (and accordingly of time), but in the Critical period he substantiates them differently. If in his earlier phase he had denied God creative functions ("give me matter and I will construct a world out of it"), later on he continues this "plundering," so to speak, divesting the deity of His sensorium by transforming space and time into the sensorium of mankind. He developed an interest in the role of theory in comprehending reality, devoting to this question his *Critique of Pure Reason*, which begins with the transcendental aesthetic and sets forth his theory of space and time.

Kant's space is not an empirical concept, as it is not derived from external experience. Moreover, the possibility of external experience derives from the representation of space as a necessary condition for the spatial location of external coexistent objects (phenomena): "Space is a necessary *a priori* representation, which underlies all outer intuitions."[196] He especially emphasizes that "space is not a discursive, or, as we say, general concept of relations in things in general, but a pure intuition."[197] There exists but one space, and

> if we speak of diverse spaces, we mean thereby only parts of one and the same unique space. Secondly, these parts cannot precede the one all-embracing space, as being, as it were, constituents out of which it can be composed; on the contrary, they can be thought of only as *in* it.[198]

It is with reference to just such considerations that Kant maintains that space should be called not *compositum* but *totum*.

Nor is time an empirical notion derived from experience.

Kant regards time as well as an a priori necessary representation underlying sense intuition. Time is a one-dimensional variety whose parts exist successively. The successiveness of the existence of parts of a single time is not a proposition derived from some general concept; instead it is synthetic and immediately contained in the intuition of time.[199]

Space and time are subjective and necessary conditions of sensory experience, like a priori forms of intuition, and this defines their empirical reality only with respect to phenomena. They have no relation to *noumena,* or things-in-themselves, which serve as the extrarational and extrasensory basis of phenomena. When the problem is thus stated, of course, *noumena* are absolutely unknowable, extraspatial, and extratemporal. Reichenbach has correctly perceived Kant's affinity with the Eleatics:

The Parmenidean distinction between timeless Being and illusory time flow reappears in Kant's philosophy as the distinction between timeless things-in-themselves and time as the subjective form of things-of-appearance.[200]

It deserves to be remarked in this connection that the world of *noumena* must be a kind of analogy with the One of the Eleatics, because plurality is inseparably associated with space and time. Significantly, on the level of reason Kant arrives at the notion of a world that is absolute totality.

Kant's theory of space and time synthesizes certain aspects of the substantial and relational concepts. As a priori forms of intuition, space and time are, as it were, coordinating factors that represent phenomena in a spatial and temporal form. That is, they are a system of relations and in this respect are indebted to Leibniz's theories. As Bertrand Russell rightly observes, however, "Kant's space . . . is absolute, like Newton's, and not merely a system of relations."[201]

In Kant's system one can speak not only of space and time as synthesizing certain features of the substantial and relational notions but also of two spaces and times. Thus T. E. Wilkerson observes that Kant operates with two temporal and spatial representations, which at times appear as pure intuitions and at other times as pure forms of intuition.[202] This fluctuation depends on whether Kant is discussing mathematics or is rather engaged in setting out the necessary conditions of experience. This results in two images of space and time—pure intuitions and pure concepts. It deserves to be noted in this connection that Kant was concerned with how categories and appearances are interrelated. He realized that there must exist some third entity that is homogeneous with both, and it is this intermediary link that allows categories to be applied to appearances. Such is the transcendental schema in which intellectual and sensible aspects are united. Specifically,

> an application of the category to appearances becomes possible by means of the transcendental determination of time, which, as the schema of the concepts of understanding, mediates the subsumption of the appearances under the category.[203]

It follows from the uniqueness of Kantian space and time that they are universal—the world of phenomena represents against a single spatial and temporal background. In this conception geometry serves as the system of synthetic a priori knowledge. First, the intuitive a priori knowledge of geometric truths (such as axioms) is the criterion of their absolute authenticity. Second, geometric axioms are synthetic judgments, that is, judgments in which the predicates add something to the concept of the subject.

No analysis of Kant's spatiotemporal concepts would be complete without a discussion of his antinomies. The first antinomy treated in the transcendental antithetic is:

Thesis: The world has a beginning in time and is also limited as regards space.
Antithesis: The world has no beginning and no limits in space; it is infinite as regards both time and space.[204]

In this (mathematical) antinomy Kant is trying to show that both opposite dialectical statements are false if we attempt to apply the idea of absolute totality (of the world) to appearances. The proof of the statements in the antinomy is constructed as follows. On the one hand, if we

> assume that the world has no beginning in time, then up to each given moment of time an eternity has elapsed and there has passed away in the world an infinite series of successive states of things. Now the infinity of the series consists in the fact that it can never be completed through successive synthesis.[205]

From this it is concluded that an infinite temporal series is impossible; that is, there must be a beginning in time. An analogous proof is presented for the limitation of the world in space—an infinite aggregate of real things cannot be regarded as a given whole.

On the other hand, the world cannot have a beginning in time and boundaries in space, because in such a model "empty time and empty space must constitute the limit of the world."[206] Here we are confronted with the question of beginnings, for how can empty time, in which no beginning is possible, be interrupted, and how can there occur such a unique event as the creation of the world?

The logical rigor of Kant's argumentation is open to question. Already Hegel noted that the premises of the antinomy involve *petitio principii*, assuming as proved that which is to be proved. As E. M. Chudinov has pointed out, a beginning in time is presumed in Kant's notion that an infinite temporal series concluded by the present moment is not infinite.[207]

Kant regarded this antinomy as an important confirmation of his notion of space and time as a priori forms of sense intuition. The negative answer to both alternatives indicated that space and time are nonobjective, that these forms cannot be applied to the world as an absolute totality, and that they are merely subjective forms of intuition.

In his comments on Kant's antinomies, Hegel wrote that "they, more than anything else, brought about the downfall of previous metaphysics and can be regarded as the main transition into more recent philosophy."[208] These antinomies also figured prominently in the dialectic of Hegel, whose works represent the culmination of classical German philosophy.

Hegel's spatiotemporal conceptions may seem unoriginal, for in his writing one often finds the views of his predecessors expressed in more cumbersome language. For example, he sets forth the relationship between duration and time as follows:

It is the universality of these present moments which *lasts,* and the sublatedness of this process of things which does not. Even if things endure, time does not rest, but continues to pass, and it is because of this that it appears to be distinct and independent of things. If we say that time continues to pass even if things endure, however, we are merely saying that although some things endure, change appears in other things, as for example the course of the sun; so that things still remain in time.[209]

To this are also appended traditional arguments on eternity as absolute timelessness, as duration reflected in itself, and so on. All the same, it is difficult to exaggerate the significance of Hegel's contribution to the evolution of spatial and temporal notions.

The following conclusion can be drawn from my analysis of pre-Hegelian philosophy: Whether space and time are regarded as empty or filled, whether space is considered in connection with or apart from motion, whether its nature is thought to be objective or subjective, the dominant approach throughout is an abstract mathematical one that treats the properties of space and time but not their structure. An understanding of structure can be gained in Hegel's approach if, that is, space and time are regarded as internal elements of the motion of matter, whose structure changes relative to the changing nature of motion itself. Hegel was the first to point out that in motion there is no one point of space corresponding to a given moment or "now" of time. As he wrote in *The Philosophy of Nature:*

> Zeno's antinomy is insoluble, and motion falls into it if places are isolated as spatial points, and moments of time as points of time. The solution of the antinomy, i.e. motion, can only be grasped through the inherent continuity of space and time, and the simultaneity of the autonomous body's both being and not-being in the same place, so that it is simultaneously in another.[210]

Corresponding to a given moment is not a point of space but a segment. The dynamic atomism of space and time is such that the greater the rate of motion, the greater the distance between two points of space occupied by a moving body at one and the same moment of time.

Hegel's dialectical theory of space and time was highly regarded by the founders of Marxism-Leninism and was adopted by materialist dialectics, which is the logic and methodology of modern natural science. Having thus nearly reached dialectical materialism in this study, in the next chapter I undertake a dialectical-materialist analysis of the evolution, status, and present developmental trends of spatial and temporal conceptions in modern physics.

3

A Philosophical Analysis of Space and Time in Physical Theory

The preceding chapters discussed the sources and genesis of spatial and temporal notions and their evolution in mythological and philosophical systems. To appreciate the true significance of these concepts, however, we must analyze their status in the structure of physical theory. Such an approach will enable us to regard space and time in their logical, epistemological, semantic, empirical, theoretical, and other aspects and to indicate the direction in which these structures are developing in modern physics.

The Status of Space and Time in the Early History of Classical Mechanics

The status of space and time in the structure of physical theory can be fruitfully approached through a rational reconstruction of the evolution of physics, an undertaking that is quite in keeping with the spirit of the present study. I am therefore obliged to return ever so briefly to ancient Greece, beginning the discussion with an analysis of Euclidean geometry, which Einstein calls the oldest branch of physics and which Imre Lakatos once described as a cosmological theory.

In his *Elements* Euclid presented an axiomatic reconstruction of ancient geometry, whose constructions were often physical in nature. His system was an informal one that was immediately related to the real space of the macroworld. It is interesting to note that, although it was intended to apply primarily to mathematics, Aristotle's theory of proof advanced in *Posterior Analytics* distinguishes various levels of mathematics, not all of which satisfy the demands of ideal deductive science. Thus arithmetic is held to be a more perfect science than geometry because

126   Chapter 3

one science is more certain than another . . . if it depends on fewer items and the other on an additional posit (e.g. arithmetic and geometry). (I mean by on an additional posit, e.g. a unit is a positionless reality, and a point is a reality having position—the latter depends on an additional posit.)[1]

Already at this level we find an increase in the number of principles that is not only quantitative but also a certain complication of basic concepts and propositions. Besides definitions and axioms, for example, Euclid's geometry also contains postulates, which are not found in arithmetic. According to S. A. Yanovskaya, this expansion is not due only to the greater complexity of geometry as compared with arithmetic; there is also the fact that Euclidean geometry is a geometry of the idealized compass and straightedge, and its algorithms are relative (algorithms of reducibility) rather than absolute, as in arithmetic.[2] In the postulates, Euclid formulated problems that he regarded as already solved. Taken into consideration were the necessary instruments (compass and straightedge) and operations that, incidentally, might also be unrealizable in practice—this is a kind of protoimage of the Gedankenexperiment. The remaining problems of construction were solved through the algorithm of reducibility.

Because it is a division of pure mathematics, arithmetic has no instruments and no material base, and it utilizes a minimum number of principles and operates with absolute algorithms. Euclidean geometry is an axiomatic deductive system, but problems of construction are solved on the reducibility algorithm, and the system implicitly contains an idealized set of instruments. It is a geometry of the compass and straightedge. Ideally, physical theories can also be described as consisting of two fundamental parts—an axiomatic system and a world of instruments (the empirical basis)—connected by operational rules. Geometric algorithms of reducibility are the forerunners of the operational rules of physical theory.

Physics and mathematics are united in Euclid's geometry, and the development of each of these aspects resulted in immense advances, such as Newton's *Principia* and Hilbert's *Foundations*. The divorce of mathematics from physics in geometry was insignificant before Hilbert, as is interestingly manifested in the well-known Kantian notion of geometry as a system of synthetic a priori knowledge.[3] It may be stated with certain reservations that, given the level of science at the time, Kant was right. The geometric axioms he discussed appeared a priori and synthetic, owing to the fact that geometry was not split into the mathematical and the physical. Nevertheless, as Rudolf Carnap observes, "Mathematical geometry is a priori. Physical geometry is synthetic."[4]

The state of the problem today is as follows. On the one hand, there is mathematical geometry as the theory of a logical structure—a deductive system about a structure is constructed on a group of formal axioms. This system gives us no immediate data about the real world. It is the application of mathematical geometry to the real world that makes up the content of physical geometry. Interpreting pure geometry empirically proves to be a rather complicated matter, because it is only through physical theory that geometry is connected with the real world. We cannot obtain separate observational verification of any geometry but must always operate with geometry (G) and physics (P) in combination. Only G + P supplies empirically verifiable statements. Here, of course, it must be remembered that the very split of physical theory into the components G + P is nontrivial.[5]

We must also bear in mind that the sharp dichotomy between the analytic and the synthetic that is characteristic of logical positivism is an oversimplification of the real course of development and can be regarded as an extreme case; many scholars, such as W. Quine and M. Black, have expressed their dissatisfaction with it. As Nicolas Bourbaki notes, there is an intimate connection between experimental phenomena and mathematical structures, even if the profound reasons for these ties are not always known. It is understandable, therefore, that so much attention should have been given to basing geometry on a union of physical and mathematical aspects. In Gonseth's epistemological conception, for example, the essence of the geometric is described as a dialectical synthesis of structure (pure relations) and its substance, or physical base.[6] In such an approach, "the geometrical is a concrete method for the realization of abstract relations that clearly expresses not only the axiomatic but also the empirical content of geometry."[7]

Taking the dichotomy between the analytic and the synthetic into account, Euclidean geometry can be regarded as the first physical axiomatic theory. Significantly, this current, so relevant to modern science, began with the axiomatic theory of space. Emphasizing this point, Einstein observed:

> The first to arise was the doctrine of spatial relations between bodies irrespective of changes in time—Euclidean geometry. It is to the immortal credit of the ancient Greeks to have created this first logical system of concepts concerned with the behavior of natural objects.[8]

Einstein quite correctly perceives that Euclidean geometry deals with the behavior of natural objects, but this behavior is intrinsically associated with notions of time and motion. These concepts are explicitly absent from Euclid's geometric constructions, although a certain kinematism is not entirely alien to them. For example, motion (superposition) is applied in the proof of one case of triangle congruence. Euclid also developed another approach to the analysis of space that was based not on the idealized properties and displacements of a body but on light rays propagated in straight lines over great distances. This is the first thesis of his *Optics*, in the proofs of which the concept of motion is more broadly represented.

Thus already Euclid's geometry contained a hidden kinematic tendency that greatly influenced the subsequent development of mathematics and physics. In calculating infinitesimals, for example, lines were often represented not algebraically and not even geometrically but kinematically; this was based not only on intuited continuity but on a kind of analog to the experimental conception of magnitudes that change in time.[9] Within geometry Felix Klein developed his famous Erlangen program, in which it was demonstrated that the content of geometry consists in studying those properties of figures that remain unchanged through all manner of movements. Finally, the similar development in physics can be described as a process by which the Euclidean axiomatics of physical space was gradually supplemented with new concepts and axioms. Along this path we arrive by stages at the other set of *Principia* created by the genius of Isaac Newton. This rational reconstruction of the genesis of classical mechanics permits us to ascertain clearly the basic status of space and time.

The first expansion of Euclidean axiomatics is connected with the introduction of the concept of time. Kinematics, which makes the description of motion possible, is based on a definition of geometry and time. In the kinematic approach the notions of motion or rest are meaningful only in relation to a rigid body, known as the frame of reference. In classical kinematics it is possible to define time with a certain degree of arbitrariness. Thus, if t is time and $\tau = f(t)$, where f is a continuous, monotonically increasing function in the interval $(-\infty, +\infty)$, then τ is also time.

Later development is associated with the introduction of the notion of mass in the branch of mechanics known as kinetics. This represents the first step from the description of motion to its prediction, and it brings us directly to dynamics, which needed the concept of force and a corresponding expansion of axiomatics. Dynamics refines the notion of time and sets limits on the arbitrariness with which it can be described: If t is time (the absolute measure of time), then any other time τ is defined by the expression $\tau = at + b$, where a and b are constants and $a > 0$.

In the schematic reconstruction of the developmental path from Euclidean geometry to Newtonian mechanics, space and time are the basis and can be described in terms of the following axioms:

1. The space E^3 is a Euclidean three-dimensional differentiable manifold.
2. The time T is an interval of the real numerical axis, and each member t of the set T is a moment of time. T is the (partially) ordered relation "earlier than" or "simultaneous with."

Various modifications of these axioms are employed in the logical reconstruction of classical mechanics. In the first serious attempt to axiomatize classical mechanics by G. Hamel, for example, the first axiom is: "There exists a real, continuous variable t—absolute time."[10]

Ascertaining the basic status of space and time in the logical reconstruction of classical mechanics far from exhausts the question of concern to us here, for it leaves unexplained many physical and conceptual aspects of the problem. The identification of time with a real variable means that values derived from time (speed, impulse, etc.) will also be identified with real numbers. In that case the mechanical equations as well will be equations not of physical values but of their images in isomorphic mappings onto a set of real values. Physical equations, however, connect not numerical values but the physical values themselves.[11]

In addition to a logical reconstruction, therefore, we also need dynamic, conceptual, and other reconstructions that together can give us an idea of the status of space and time in the structure of classical mechanics. The split of physical theory into an axiomatic deductive system and an empirical basis or set of operational rules is typical of logical reconstruction, which is possible only in the case of simple theories, such as the classical mechanics of the particle (McKinsley, Sugar, Noll, and others). Such a structure is more or less an ideal. This form of logical reconstruction of physical theory, of course, is consistent with contemporary formal and formalized mathematical axiomatization. Newtonian mechanics, although it is axiomatic and can be regarded as a logical reconstruction and generalization of the work of numerous predecessors, such as Copernicus, Galileo, Descartes, and Kepler, is nevertheless an example of an informal axiomatics that abounds in drawings and in which the structural components of physical theory are in immediate unity. The latter fact derives in many respects from the direct connection and degree of isomorphism that exist between the theoretical and observational worlds of Newtonian mechanics. The ontologized theoretical state of affairs coincides with the idealized observational one. Naturally, this does not imply that there is no division in the mechanics into theoretical and empirical levels, and I devote special attention to this question later.

Between the physical informal axiomatics of Newtonian mechanics and the formalized logical axiomatics of classical mechanics lies a considerable number of logicomathematical reconstructions. There we find that constructions become increasingly rigorous, that many physical concepts are explicated, and that the number of principles is minimized. One example is the numerous elaborations by the eighteenth-century mathematicians Euler and d'Alambert, who independently aimed at a logicomathematical reconstruction of Newton's mechanics. These attempts were brought to relative completion by Lagrange,[12] in whose analytical mechanics the geometrism of Newton's *Principia* was completely eliminated and mechanics became a division of analysis. Later undertakings included Lie's group-theoretical and Wheeler's topological reconstructions.

This process gradually brought to the fore such notions as laws of conservation, principles of symmetry, and invariance, all of which made it possible to view classical physics from unified conceptual positions. These notions proved heuristic in the construction of modern (relativistic and quantum) theory and also helped elucidate important aspects of the status of space and time in the structure of physical theory.

Conservation laws, of course, were known before Newton and were regarded as principles or postulates. After Newton, however, they begin to play a different role in physics, serving as theorems derived from the universal axioms of dynamics. Lagrange's analytical mechanics established a connection between the basic laws of conservation and spatiotemporal symmetry, a question that has been comprehensively studied and developed by Lie, Klein, and Noether.[13] The conservation of such fundamental physical magnitudes as energy, momentum, and quantity of motion proved to derive from the isotropism and homogeneity of space and time. Ford notes:

Three of the seven absolute conservation laws arise solely because empty space has no distinguishing characteristics and is everywhere equally empty and equally undistinguished.[14]

Here it is interesting to observe that the ancient Greeks' postulation of empty space brought them to positions close to Newtonian dynamics. Thus Epicurus wrote Herodotus that "the atoms must move with equal speed, when they are borne onwards through the void, nothing colliding with them."[15] This physical isotachys (there was mathematical isotachys as well—a corollary of the discreteness of space and time at the level of the *amer*) anticipates Newton's first law of mechanics, which states that a body will preserve its state of rest or uniform rectilinear motion so long as no force affects it from without. On the other hand, Aristotle's rejection of the void led him to a superficial dynamic scheme that held that a body moves uniformly under the influence of a constant force.[16] The revolution in modern physics is connected with the refutation of this concept and of the Aristotelian notion of natural places, cosmic spheres, etc.

It was discovered in classical mechanics that fundamental physical laws are determined by the symmetry of space and time. Thus the law of conservation of momentum is based on the homogeneousness of space; the law applying to the conservation of quantity of motion is due to the isotrophy of space; and the homogeneity of time determines the law of conservation of energy. The symmetry of space and time accounts for the invariance of physical laws relative to certain transformations. The equations of Newtonian dynamics, for example, are invariant with respect to rotations and parallel displacements. As E. P. Wigner emphasizes, there is a profound analogy between the relationship of natural laws to phenomena on the one hand and the relationship of principles of symmetry to natural laws on the other.[17] If the laws of nature define a structure or interconnection in the world of phenomena, the principles of symmetry structure these laws by establishing the internal connections between them. It would seem that Wigner's argument can be transformed into the following thesis: Natural laws determine the structure of the observational world, whereas principles of symmetry determine that of the theoretical one. (This, of course, should not be construed to mean that principles of symmetry cannot operate at the observational level, because in a group-theoretical approach, for example, the group of symmetries that generates the means by which investigated objects can be identified with each other coincides with the symmetry groups of the state of the apparatus.) In classical physics there are geometric principles of invariance that are connected with the symmetry of space and time.

Not only are space and time the basic concepts of classical mechanics, then, but they are also the ordering structures of theory. Their status as such was discovered in the various equivalent formalisms (logicomathematical reconstructions of Newton's dynamics) developed in the eighteenth and nineteenth centuries. These formalisms were based on different fundamental mechanical principles, but they all attempted to arrive at the basic mechanical laws of conservation (that is, to derive the first integrals of mechanics). This helped to explain various aspects of the structure of classical mechanics.

The idea of generating different classical mechanics through a logicomathematical reconstruction of Newtonian mechanics can be retrospectively extrapolated onto pre-Newtonian systems as well, which may prove useful in investigating certain aspects of the dynamic and conceptual status of space and time. Instead of the series "geometry-kinematics-kinetics-dynamics," this approach deals with the chain "Aristotelian dynamics–Galilean dynamics–Newtonian dynamics."

In my discussion of the advance from Euclidean geometry to

classical mechanics, I focused attention mainly on the expansion of Euclidean axiomatics. The metric structure of space remained more or less fixed, although the definition of time underwent some refinement. A more precise image of the world emerged as its various geometric, kinematic, and dynamic sections became interrelated and commensurate. Some elements of the dynamic and conceptual status of space and time were only implicit in this process, of course, although they are clearly present in Newton's *Principia*. As regards the rational reconstruction of their genesis, this can be obtained through an analysis of the stated series of dynamics.

Considerable scholarly attention has been given to such a reconstruction of dynamics, which is immediately related to the Erlangen approach to the genesis of physics as applied by N. P. Konopolev, G. A. Zaytsev, R. Penrose, and others. This reconstruction derives its fruitfulness from the relationship of space and time in the theory of relativity; this relationship must be described in the four-dimensional language of space-time. In fact, we encounter an analysis of geometry in this reconstruction because dynamics itself becomes a geometric subject. This is the approach taken by Roger Penrose in his analysis and reconstruction of dynamic theories, especially those of Aristotle, Galileo, and Newton. Each of these dynamics has its own space-time, and although all are smooth four-dimensional manifolds, each is assigned some additional geometric structure that reflects a characteristic dynamic aspect.[18]

Aristotle's space-time is simply the product of $E^3 \times E^1$, where the metric E^3 describes spatial separations and the metric E^1 describes time segments. In Aristotle's dynamics one can speak of an absolute distance between two events in space, even if the temporal difference between them is not equal to zero. Such a position is meaningless in the space-time of Galileo and Newton. As Penrose notes,[19] the structure of geometry can thus be compared with a space fibered along E^1 with E^3 as fibers, so that the "time" E^1 can be understood as a factor space of full space relative to the fibers E^3. The topology in these space-times remains the same as in Aristotle's dynamics, but the structure uniting the fibers is different.

The structural difference between the space-times of Galileo and Newton can be explained by means of geodesic curves, which are world lines of particles moving by inertia. To discover a family of such curves in Galileo's dynamics, we need only postulate them as a system of straight lines in some E^4, where the fibers of E^3 are identified with the maximum system of mutually parallel planes in E^5 and the fiber E^1 is defined as usual as a factor space. Galileo's space-time is a particular case corresponding to a zero-point (or everywhere constant) gravitational field. This is the space-time of classical mechanics. Newton's space-time generalizes this "plane" structure as a curved one arising in a heterogeneous field. The equation of the field of Newton's space-time is obtained from the equations of the general theory of relativity in the passage to the limit $c \to \infty$.

This dynamic reconstruction of classical mechanics enables us to trace the gradually increasing complexity of space-time. In contrast to the reconstruction of the path from Euclidean geometry to Newtonian mechanics, which complicated and refined the structure of a single world, here we have a number of different worlds, each of which is an interpreted structure of a corresponding space-time. The status of space and time is no longer simply basic—space and time unite in the integral structure of space-time, which defines dynamics and its corresponding groups of movements. This current reaches its logical conclusion in the characteristic thesis of geometrodynamics (Wheeler, Misner) that there is nothing at all in the universe but curved space.

Returning now to Penrose's structures of dynamics, it should be observed that, although a retrospective reconstruction of physical theory can teach us a great deal about the structural aspects of the reconstruction, we are often operating in another picture of the world and with a different conceptual apparatus. Thus, in the passage to the limit $c \to \infty$, the formalism of the special theory of relativity can be combined with the scheme of classical mechanics, or the field equations of the general theory of relativity can be reduced to equations of the Newtonian theory of gravitation; in doing so, however, we are not entering directly into the world of classical physics but remaining within a limited fragment of Einstein's universe. The tendency of a theoretical parameter to approach a limiting value is not automatically accompanied by conceptual changes.

I therefore consider the spatial and temporal conceptions of classical mechanics, as only such an analysis can provide a complete picture of the status of space and time within that theory. It is not enough to define the notions and axioms of space and time or to reconstruct the dynamic structure of space-time.

Other factors that must be taken into account include the conception of space and time, the world image associated with it, the interrelation of theory and the observational level, and the characteristics of the operational rules. Determining the status of space and time in physical theory demands that all these components be viewed together.

Let us proceed directly to Newton's *Principia*, which is presented axiomatically, and see where and how its author defines the notion of space and time. *Principia* begins with a definition of basic physical concepts, such as mass, quantity of motion, inertia, and force. This is followed by the concepts of absolute and relative space, time, and motion, to which is dedicated the precept concluding the first chapter. The second chapter presents the three laws of motion as axioms.

Thus spatial and temporal notions are introduced at the level of primary terms and derive their physical meaning from axioms or laws of motion. They precede the axioms not only because they are defined by them but also because they prescribe the background on which the axioms themselves are realized. These laws of motion in classical mechanics hold in inertial reference frames, which are defined as systems moving inertially relative to absolute space and time.

At this stage the primary theoretical status of absolute space and time—the "box without sides" and pure duration—becomes evident. This is reflected in the well-known thesis of Newton's *Principia:*

Absolute, true, and mathematical time, of itself, and from its own nature, always flows equably, without relation to anything external, and by another name is called duration. . . . Absolute space, in its own nature, without relation to anything external, remains always similar and immovable.[20]

Newton's absolute space is an analog of the Democritian void, and Newton could not have developed his atomic doctrine apart from the fundamental thesis regarding the two principles—atoms and the vacuum. The Democritian void was the basis of one attributive notion—the extensional, which views space as an extension of material objects. The other attributive concept of the space of material objects is known as relational.[21]

For Newton, absolute space and time were the arena in which the dynamics of physical objects acted. In contrast to Democritus's void, however, Newtonian space is bound up with a certain mathematically shaped dynamics and is given physical meaning by laws of motion, whereas its symmetry accounts for the fundamental laws of conservation. As Einstein observed, this space "is assigned an absolute role in the whole causal structure of the theory."[22]

As many critics of the Newtonian conception have pointed out, the definition of absolute space is in many respects contradictory. Thus Philipp Frank holds that the Newtonian physics lost logical coherence as soon as attempts were made to rid it of theological elements in the late eighteenth and early nineteenth centuries. Until that "purge," so long as absolute space was regarded as the *sensorium Dei*, everything was in order and Newton's physics was logically consistent. Furthermore, following Mach, Frank suggests that absolute space lacks operational meaning and therefore becomes merely an empty entity. To get rid of this meaningless term, he considers the following two premises: (1) The law of inertia is valid with respect to absolute space; (2) fixed stars are in a state of rest with respect to absolute space.[23]

From this it is concluded that the law of inertia is valid with respect to fixed stars. It is fine, of course, to be able to find an empirical protoimage of absolute space (or rather a fairly well-fixed inertial reference frame), but it is unclear why this should lead to the conclusion that absolute space is to be excluded from classical mechanics and its laws. Fixed stars are of practical value in empirical investigations, but they cannot serve as a theoretical structure (a structure of the theoretical world)—the fundamental laws of conservation, etc. are defined not by fixed stars but by the symmetries of absolute space.[24]

All such problems concerning the empirical over the theoretical, the observable in contrast to the unobservable, and so on are immensely significant in modern physics. As regards the definition of absolute space, Einstein observes that on this point Newton was particularly consistent:

[Newton] had realized that observable geometrical magnitudes (distances of material points from one another) and their course in time do not completely characterize motion in its physical aspects. He proved this in the famous experiment with the rotating vessel of water. Therefore, in addition to masses and temporally variable distances, there must be something else that determines motion. That "something" he takes to be relation to "absolute space."[25]

Absolute space and time are thus the necessary theoretical basis of classical mechanics.

Corresponding to absolute time, classical mechanics also postulated absolute and universal simultaneity. Only instantaneous forces acting at a distance can serve as the basis of absolute synchronism, and the role of such forces was assigned to gravitation (the universal law of gravitation). Actually, this feature is found already in classical kinematics, in which the concept of time rested on the following hypothesis:

> Two events which are simultaneous for an observer bound to a given bench mark [frame of reference—M. A.] will appear equally simultaneous to an observer bound to an arbitrary bench mark moving relative to the first one.[26]

Such a correspondence is possible physically, given signals traveling at infinite velocity. Thus action-at-a-distance was not generated by Newton's dynamics but was present already in kinematics. And even if Newton had succeeded in constructing a near-action theory of gravitation within an ethereal model (he did not approve of action-at-a-distance and in fact attempted to develop ethereal models), either he would have had to find some substitute that acted at a distance or he would have been obliged to anticipate Lorentz, Poincaré, and Einstein by undertaking a relativistic reorganization of mechanics. The status of action-at-a-distance is not dependent on the nature of gravitation but derives from the substantial conception of space and time intrinsic to the mechanistic picture of the world. Within that view, assuming the existence in nature of a *maximum* and a *minimum* velocity, c results in the illogical expression $c + v = c$ if $v \neq 0$. It becomes logical only on the condition that $c = \infty$. Extratemporal and extraspatial instantaneous action-at-a-distance were the logical framework of absolute space and time.

From absolute space Newton distinguishes the extension of material objects, which is both their basic property and relative space (this is the extensional conception). It is important to emphasize that relative space is the measure of absolute space and can be described as a set of concrete inertial reference frames in relative motion. Relative time is accordingly also the measure of duration, which as hours, days, months, and years is used in everyday life instead of true mathematical absolute time. Relative space and time are accessible to the senses.[27] One occasionally encounters the view that Newton's relative space and time are the private visual space and subjective psychological time of immediate sensory experience.[28] This contention has been justly criticized by Adolf Grünbaum, who rather persuasively shows that relative space and time are empirical rather than perceptual. He emphasizes that Newton's

> relative space and time are indeed that public space and time which is defined by the system of relations between material bodies and events [that is, here is meant not even the extensional notion but a relational variant of the attributive conception of space and time—M. A.] and not the egocentrically private space and time of phenomenal experience.[29]

Thus, besides theoretical space and time, which are defined by mechanical laws and (in Newton's own terms) are mathematical, Newton also introduces empirical space and time, which are sense perceptible, serve as the measure of theoretical structures, are used in everyday life, and are described in the language of observation.[30] Empirical verification and interpretation occur within the framework of this empirical (relative) space and time.

Status and Directions of Spatial and Temporal Conceptions in Modern Physics: Theoretical and Empirical Structures
The Possible Macroscopic Nature of Space and Time

We have seen that in Newtonian mechanics theoretical and empirical space and time coincide to a considerable degree. I henceforth refer to them as the T- and E-structures of physical theories, as they are a particular case of the theoretical (T) and empirical (E) structures of relations in such theories. Newton in particular emphasized that "absolute and relative space are the same in figure and magnitude."[31] This is natural because both structures refer to the same macroworld and to changes in its objects. At the level of Newtonian mechanics, of course, we do not find the identity of theory and observation insisted on by extreme empiricists, such as John Stuart Mill, but the connection between these spheres (and thus between the T- and E-structures) is immediate, and they are "written" in the same spatiotemporal language. It should be noted here that the objects of both the theory and the empirical basis are to a certain extent idealized, albeit at different levels.

Theory does not describe real phenomena themselves but their idealized analogs. Nor are the empirical objects of theory simply sense perceptions. Objects of the observational level are also the products of rationally processed sense experience. Scientific methodology long ago rejected the theory-independent Protokollsätze developed by the Vienna Circle in the 1920s and 1930s. Empirical objects bear the stamp of theory (the so-called theoretical load of the observational language). All this notwithstanding, objects on the empirical level are more immediately connected with objective reality and are in a sense "given" to us. For that reason, therefore, verifying the adequacy or correspondence to reality of a theory involves correlating theoretical constructs with empirical data. Without such correlation the theory loses its relationship to reality. Correlation is attained by means of special correspondence rules whose core in developed scientific disciplines consists of measurement procedures, which are among the most important structural elements of physical theory. As Carnap has observed, the measurement of length and temporal duration elevates them to the level of fundamental magnitudes in physics:

Once we can measure them, many other magnitudes can be defined. It may not be possible to define these other magnitudes explicitly, but at least they can be introduced by operational rules that make use of the concepts of distance in space or time.[32]

Such definitions of physical magnitudes have led to the far-reaching conclusion that spatiotemporal terminology is capable of expressing all aspects of physical investigation—physical knowledge can be expressed in a coordinate language by means of such terms as "space," "time," and "point."

The universality and fundamental nature of the notions "length" and "interval of duration" were absolutized in the logical positivism of Carnap, Hempel, and Russell. This gave rise to numerous difficulties, which the positivists attempted to solve by means of reductional propositions, Carnap's S-rules, the elimination of theoretical terms according to Ramsey, etc.[33] I do not analyze here the features of "unobservable" theoretical objects or consider nomological statements devoid of operational meaning. The relationship of such elements of a theory to its empirical structure is ambiguous; moreover, "unobservables" may not have any spatiotemporal meaning at all. Instead,

I attempt to touch on another aspect of the problem by examining the "divorce," or polysemy, of basic concepts that led to the rejection of a unified empirical basis among different physical theories in favor of the view that these theories have distinct E-structures and that the observational language bears the imprint of the corresponding theory. (Typical in this regard is Einstein's well-known thesis that only theory can determine what is or is not observable.)

Riemann once inquired as to the possible macroscopic nature of the Euclidian space and time with which classical physics operates. He observed:

The empirical notions upon which the determination of spatial metric relations is based—the rigid body and the light ray—seem to lose their validity in the infinitely small. It is therefore quite conceivable that the metric relations of space in the infinitely small do not correspond to geometrical assumptions, and we should in fact accept this if it can help us to explain observed phenomena.[34]

This passage suggests, first, that the time and space of classical physics may be macroscopic; this means that its operational procedures will change in the microworld. But there is also one thesis here that describes an important feature of classical physics, and revision of it defines another way in which classical operational procedures change. Here I mean the single-order, isotheoretical use of the *rigid body* and the *light ray* for the physicalization and metrication of the space of classical physics. The significant feature underlying this in classical physics is that mechanical and optical processes are unified because optics is a branch of mechanics. Classical mechanics has a mechanical operational level. At first it made do with the straightedge, the compass, the pendulum, etc.; that is, it had a geometromechcanical operational base. Subsequent astronomical studies demonstrated the superiority of optical processes, and this base began to become opticomechanical in nature. (It was William Rowan Hamilton's "astronomical" approach, for example, that led him to the theory of optical instruments and the opticomechanical analogy.) Methods and ideas of optics, such as the principle of least action, began to find their way into mechanics. Now the operational level exerts a corrective influence on the basic theory: *The theory of the operational level, that is, optics, corrects the fundamental theory, that is, mechanics.* Optics, although it was regarded mechanically (at a time when even nonphysical objects and processes, such as society, were regarded like this, it would have been strange to find physical and particularly optical phenomena considered in any other way), nonetheless possessed certain features that influenced the development of mechanics, for example, the principle of least resistance. As physics advanced, however, these features proved to be less and less in conformity with the conceptual apparatus of mechanics.

The development of wave optics, for example, led to significant shifts in spatial and temporal notions because "light waves were, after all, nothing more than undulatory states of empty space, and space thus gave up its passive role as a mere stage for physical events."[35] Various ethereal palliatives were proposed to rescue the mechanical world view, but Maxwell and Lorentz demonstrated that electrodynamics cannot be reduced to mechanics. In field theory the physical state of space itself is a physical reality. Deserving special emphasis here is the fact that light belongs in the realm of electromagnetism!

Thus, if physicists wanted to be consistent, they somehow had to consider that, if optical processes are used in the operational procedures of classical mechanics, then that mechanics will have an electromagnetic operational level. Ignoring this point was bound to give rise to paradoxes in theses concerning near–light velocities, and thus the question was also raised of whether the electrodynamics of moving bodies was developing consistently. This entire range of problems became the central concern of physics around the turn of the present century, and it was to these issues that such outstanding thinkers as Lorentz, Poincaré, and Einstein turned their attention.[36]

The operational procedures used to physicalize Euclidean space in classical physics proved inapplicable in the region $v \sim c$. This is why, in constructing his special theory of relativity, Einstein begins by considering definitions of simultaneity and uses a new operational procedure based on light signaling. Other considerations rest on the principle of relativity and the constant speed of light. As L. I. Mandel'shtam emphasizes:

The fact that light travels at a specific finite velocity assumed enormous significance in the theory of relativity. . . . It is as central to the theory of time as the rigid body is to the theory of space.[37]

The relational conception of space and time developed accordingly, and it was asserted not altogether correctly that that notion had replaced the substantial one in the theory of relativity. The theory of relativity, like classical mechanics, operates with two types of space and time embodying the substantial and the attributive (in this case the relational) conceptions, respectively. Absolute space and time are the T-structure in classical mechanics and represent the substantial notion. Unified four-dimensional space-time has an analogous status in the theory of relativity. Einstein describes this state of affairs quite clearly:

> Just as it was necessary from the Newtonian standpoint to make both the statements *tempus est absolutum, spatium est absolutum*, so from the standpoint of the special theory of relativity we must say, *continuum spatii et temporis est absolutum*. In this latter statement *absolutum* means not only "physically real," but also "independent in its physical properties, have a physical effect but not itself influenced by physical conditions."[38]

Nor does the general theory of relativity reject absolute space. As Grünbaum shows, to solve the theory's nonlinear differential equations, boundary conditions must be supplied "at infinity," and these then assume the role of Newton's absolute space.[39]

The world of classical physics was bisubstantial, with absolute space and time as the background and matter as the filler. Einsteinian physics steadily transformed filler components into background structures, such as curvature, torsion, and dimensional increase, giving rise to the general theory of relativity, unified field theories, and geometrodynamics. Einstein's ideas about unified field theories are symptomatic:

> We arrive at the odd conclusion that space is primary; matter is to be obtained from space at the next stage, so to speak. Space absorbs matter.[40]

Misner and Wheeler, the developers of geometrodynamics, conclude that the universe contains nothing but empty curved space: Physics is geometry.[41] In this sense geometrodynamics is an extreme manifestation of the substantial conception (monosubstantiality).

On the other hand, the relational conception of space and time is also quite obviously present in the special theory of relativity, and it was precisely this notion that took form in Einstein's first studies. As for unified four-dimensional space-time, it is a later result of Hermann Minkowski's logico-mathematical reconstruction of the theory,[42] within which space and time are the E-structure.

The theoretical and empirical components of the theory of relativity are commonly regarded as divorced. Thus A. M. Mostepanenko writes:

The four-dimensional spatio-temporal manifold is the basic theoretical object described by the theory of relativity; as for space and time separately, they become empirical objects which are essentially "projections" of unified space-time onto the corresponding frame of reference.[43]

The transition from classical mechanics to the special theory of relativity can thus be described (1) at the theoretical level (a transition from absolute and substantial space and time to absolute and substantial space-time) and (2) at the empirical level (a shift from extensional relational space and time to relational space and time). It is therefore incorrect to attempt, as certain observers have done, to prove a lack of correspondence between the notions of the old and new theories by comparing the T-structures of the old theory with the E-structures of the new. For example, A. F. Zotov claims:

Properly speaking, the new notions which qualitatively distinguish the new theory from the former one, with which they are bound by a correspondence relation, do not satisfy the correspondence principle. The concept of "absolute space" fundamental to classical physical theories is by no means the "limiting case" of the notion of "relative space" in the special theory of relativity.[44]

If we compare these concepts at the same level, however, we reach the different conclusion that the four-dimensional space of the special theory of relativity is as rigid and absolute as Newton's space. These notions correspond.

The situation arising from the expansion and generalization of T- and E-structures has become a typical feature of modern physics. Here, however, an important problem presents itself. When it was ascertained that the rigid body and the absolutely synchronized clock were inapplicable to the relativistic universe and that the rigid body and the light ray could not be used in the microworld, it became apparent that changes would have to be made not only in the operational procedures used in the study of different worlds but also in the meaning of basic concepts themselves, particularly that of length. P. W. Bridgman, who regards physical notions as synonymous with sets of experimental operations of measurement, has addressed this problem of ambiguity. Viewed as such sets, a single notion (for example, length) splits into a family of concepts. Bridgman notes of this point that "to say that a certain star is 10^5 light years distant is actually and conceptually an entirely different *kind* of thing from saying that a certain goal post is 100 meters distant."[45] Bridgman therefore attempts to establish the hierarchy of various notions of length on the basis of which operational concepts are reducible to others. Some of these concepts of length proved mutually reducible, but there are also "operationally incommensurable" ones, such as the notion of optical length in cosmic studies and the length of microworld objects.

I do not go into Bridgman's extreme operationalistic claims here. He has subsequently moderated them, conceding some value to mental, paper-and-pencil, and verbal operations. It is generally recognized that there are no grounds for absolutizing operational definitions of a physical concept. The fact that the operational split of a unified notion occurs, however, merits attention, as it reveals a theoretical-empirical dualism in the definition of the concept. Here we encounter a kind of vicious circle: On the one hand, we attempt to reduce all theoretical notions to the empirical concepts of length and duration, and in this sense space and time are the basic, empirical structures of physical theory. On the other hand, it turns out that in defining length in the microworld or megaworld we are forced to operate with concepts of length that include not merely theoretical elements but entire theories, such as the theory of electricity and Maxwell's equations. Moreover, we are obliged to extrapolate these equations to areas in which their accuracy becomes problematic.

Thus, when different physical theories are said to have different empirical bases, it must not be forgotten that notions that seem common to several theories can on closer examination prove to be distinct. This is of considerable significance in the logical problems connected with theory comparability. Thus, in considering the meaning of the difference between the paradigms of two scientific theories, Mary Hesse places the main stress on changes in the sense of terms: (1) the meaning of a term in one theory is not the same as its meaning in another;

(2) no statement, in particular no observation statement, containing the predicate in one theory can contradict a statement containing the predicate in the other theory; (3) no empirical statement belonging to one theory can be used as a test for another theory.[46] Thus no experiment can determine which of two theories should be selected.

As regards the split in notions that occurs when they are included in the empirical bases of different theories, it must be remembered that I am speaking of basic concepts, such as length and duration. The different concepts of length used in the empirical bases of various physical theories are defined by distinct operational procedures. We cannot rule out the possibility that advances in physics may oblige us to draw precise distinctions between different notions of length. Naturalists are reluctant to get involved in such concept splits. L. B. Bazhenov argues in this connection that basic concepts possess the special feature of "extrapolationability":

We could, of course, simply call mass in relativistic and classical theory by different names and regard them as distinct notions. However, owing to the feature which I have proposed be termed the conservability or extrapolationability of basic concepts, we usually refrain from doing so, and choose instead to extend the old notion to a new area and change its meaning.[47]

This is indeed what normally happens. But not always, for extrapolationability is not unlimited. We are forced to introduce conceptual splits in mathematics: If *distance* is given in real numbers, then *deviation* is defined in more general structures, such as partially ordered sets. A similar state of affairs can arise in physics, where we may find ourselves obliged to introduce various analogs of length, such as length, distance, and remoteness, together with the following logical and operational hierarchy.

The operational split of spatiotemporal concepts entails a change in the E-structure itself of a physical theory.[48] Because the E-structure always functions as a structure of the macroworld, which is the world of observational verification and interpretation, it is nevertheless regarded as a relatively stable component of the theory. Here the fact that we ourselves are macroscopic beings exerts an influence.

The question presents itself, however, of whether the E-structure of a physical theory absolutely must have the form of the space and time of classical physics. In the theory of relativity, which like classical mechanics remains a theory of the macroworld, this question is generally answered in the affirmative, although here as well we must take into account the fact that the extensional conception has been expanded to the relational one. The situation is more complicated in quantum mechanics, which is a theory of the microworld, and the founders of quantum physics have given considerable attention to this relevant problem. It was in fact in addressing it that fundamental principles, such as correspondence and complementarity, were developed. Werner Heisenberg describes the state of affairs that has arisen in microworld physics:

> It appears that a peculiar schism in our investigations of atomic processes is inevitable. On the one hand the experimental questions which we ask of nature are always formulated with the help of the plain concepts of classical physics and more especially using the concepts of time and space [the E-structure—M. A.]. For indeed, we possess only a form of speech adapted to the objects of our daily environment and capable of describing for instance the structure of some apparatus of measurement. Our experiences, too, can only be made in time and space. On the other hand, the mathematical expressions suitable for the representation of experimental reality are wave functions in multidimensional configuration spaces [the T-structure—M. A.] which allow of no easily comprehensible interpretation.[49]

It can thus be concluded that the space and time of classical physics are the empirical structure of quantum mechanics. This would also seem to follow from Bohr's famous epistemological principle, which states that "however far the phenomena transcend the scope of classical physical explanation, the account of all evidence must be expressed in classical terms."[50]

It has often been pointed out that the chief difficulties in Bohr's principle have to do with interpreting the terms "classical," "classical system," and the like.[51] What is needed is a structural and not merely a historical description of the meaning of the notion "classical." From a mathematical viewpoint the classical system can be defined as a system in which all "observables" intercommute. This does not solve the problem, however, because from the physical standpoint such systems can be realized only in an approximate sense. We must have additional criteria.

It may be concluded on the basis of Bohr's principle that classical space and time are universal.[52] Generally, however, this conclusion does not follow from the principle, because the principle is indifferent to structure type and demands merely that the structure be classical rather than quantal in nature. Thus we can say that Bohr's principle requires that the E-structure of a physical theory be macroscopic (and in this sense classical), but it does not insist that this structure be embodied in classical space and time. The E-structure of any physical theory, although it remains macroscopic (here we must recall different macroworld structures or, rather, structures of different macroworlds as presented in Penrose's reconstruction[53]), nevertheless bears the imprint of the theory itself and does not remain immutable: Observational verification and interpretation of the theory take place in the geocentric world of a macrosubject who is close to the theory itself.

Thus we can presume that the E-structure does not always appear as the space and time of classical mechanics. Modern physical theories tend instead to be associated with relativistic concepts. A. Z. Petrov holds that "the contemporary experimenter describes all physical measurements within the framework of the theory of relativity."[54] Some evidence suggests that our perception of the world is essentially and, with respect to general laws, far closer to the spirit of relativistic physics than to classical physics. David Bohm emphasizes that

nonrelativistic notions appear more natural to us than relativistic notions, mainly because of our limited and inadequate understanding of the *domain of common experience,* rather than because of any inherent inevitability of our habitual mode of apprehending this domain.[55]

As R. H. Atkin notes in his review of quantum theory and directions in its contemporary development, the scientist of the future may well be obliged to work with the topological structure of a set of observations, take into account the homology of space-time, etc.[56]

In the transition from one fundamental physical theory to another, we find significant alterations in the empirical basis, which operates in its specific space and time, in the specific E-structure of the physical theory. The process by which a given E-structure develops is a long one and often requires special reconstructions of theory. One illustrative example is quantum theory, in which the empirical basis is not fully clear despite the highly developed formalism of Hilbert space. The E-structure of this theory has been borrowed from classical physics or at best from the special theory of relativity. Many observers believe that the spatiotemporal order corresponding to quantum theory has not yet been elucidated.

This problem has been thoroughly scrutinized by such contemporary scientists as C. F. von Weizsacker, David Bohm, Ya. Aaronov, and S. Hiley. Thus Aaronov points out that we usually regard quantum theory through the prism of the "impossibility principle," focusing attention on the limitations that it introduces into classical physics.[57] What must be done, however, is to bring to light and take fuller account of a new content. Specifically, we must determine what new experimental stances are dictated to us by quantum theory as against the classical ones. Here we can expect significant changes in the informal language of quantum theory, in which such notions as particle, field, and continuum may have to be rejected to attain congruence with the formal language.

Yet, despite the great variety of E-structures (a diversity, moreover, that is bound to increase as physics develops and penetrates deeper into the structure of matter), they do share a common macroscopic nature. This is not directly related to the well-known hypothesis about the macroscopic character of space and time, because that hypothesis has to do with the theoretical rather than the empirical structure of physical theory.[58]

The point is that it is becoming difficult indeed to interpret the T-structures of modern nonclassical theories in terms of space and time. These structures are extremely complex mathematical manifolds (interpreted empirically and semantically), such as the Riemannian space of relativistic theories, the infinite-dimensional Hilbert space of quantum mechanics, the foamlike structure with a fluctuating topology of quantum geometrodynamics, and so on.

Under such circumstances the hypothetical macroscopic nature of space and time is, so to speak, up in the clouds. Indeed, in classical mechanics, which describes the dynamics of macroworld objects, the T-structure has the form of space and time. In quantum mechanics, on the other hand, which describes microworld processes, the T-structure is infinite-dimensional Hilbert space.

Here, however, we should bear in mind that, if physics had followed Poincaré, traditional classical Euclidean space and time would have remained the universal background against which bizarre physical phenomena and processes occur. The simplicity of geometry (G) would have considerably complicated the physical "filling" (P), giving rise to universal Reichenbachian forces, and so on. Physics would have taken a different path of evolution, the one traveled by Einstein: In the complex G + P that constitutes a physical theory, G is not fixed but is selected in accordance with a fairly simple and natural P. In the transition from Poincaré's notions to the situation obtaining in Einstein's theory of relativity, the very concept of simplicity was altered. This is ably shown by E. M. Chudinov:

> The regulative function of the simplicity principle was expressed in the demand for simplicity not of the geometrical component, but of the entire G + P system. Also the notion of simplicity is used here in a different sense than Poincaré's. It is of the Ockhamite variety and is intrinsic to the general theory of relativity. There G and P were combined in such a way as to guarantee that the premises of the entire theoretical system would have the greatest possible universality.[59]

What is the essence of the hypothesis that space and time are macroscopic in nature? The E-structure is of course macroscopic—here there is nothing hypothetical. In a description of the macroworld the T-structure is space and time (naturally, this does not imply that classical physics cannot be developed in another structure). In the development of physical theories describing other levels in the structure of matter, however, the use of this background, although possible, involves a complication of the theory and is therefore not generally adopted. It is this state of affairs that constitutes the simplified basis of the hypothesis on the macroscopic nature of space and time.[60]

True, there is another method that has attracted attention lately and that attempts to crystallize space-time from the specific elements of nonclassical physical theory. Yu. B. Rumer's logicomathematical analysis of this approach charts the interesting evolution of our spatiotemporal conceptions. There we see that the space-time of various physical theories at different developmental stages is connected with different basic concepts. Thus in classical physics (the first stage) these notions are energy and momentum. Owing to their basicity, the concepts of space and time associated with them began to be regarded as universals. Rumer emphasizes:

Yet a process had long been under way in physics to move beyond these concepts. Various space-like structures [that is, T-structures— M. A.] such as phase-space evidence the fact. Discoveries in the area of elementary particles associated with Gell-Mann, Neeman, Okubo and others convincingly showed that besides the space-time of momentum and energy it was necessary to introduce new types of space with new properties of symmetry. One example is three-dimensional unitary space in the SU(3) theory; in other words, the space-time connected with energy and momentum no longer has a monopoly.[61]

Thus the space introduced will organically correspond to the theory. The traditional development of quantum mechanics, for example, involved quantizing geometry, by which is not meant a shift to quantized space-time in the manner of Snyder. Instead, the geometric (and accordingly also the dynamic) structure of classical physics, which within certain limits had quite correctly described the real world, was improved by quantization. D. Finkelstein indicates yet another approach—not the quantization of geometry but the geometrization of quanta![62]

The point is that in quantum theory we have a great many magnitudes and concepts of an essentially nonclassical quantal character. Particle spin is one such parameter. Interesting attempts have been made recently to develop quantum theory solely on the basis of nonclassical magnitudes. In this approach a theory is constructed that, in accordance with the revised conception of the continuum, dispenses entirely with the concepts of particle and field. These imply that the object can potentially be isolated, a notion that is incompatible with the basic idea of wholeness on which quantum physics is based. Space and time here are not merely eliminated from the theory (as they are in the view that they are macroscopic in nature) but are constructed on the basis of nonclassical, quantum-generating elements. Space-time thus constructed possesses a great many nontrivial properties—it is combinatorial, has a spinor structure, and so on.[63] Attempts are being made along these lines to develop a better structured quantum theory in which congruence can be attained between formal and informal language.

Thus the evolution of quantum physics illustrates three ways in which nonclassical theories can develop: (1) within the framework of classical space and time—this involves unnatural complications in the theory and is sometimes employed in the early stages of theory development; (2) within the framework of abstract mathematical manifolds such as infinite-dimensional Hilbert space—this approach offers physics the means of utilizing the rich stock of abstract structures made available by modern mathematics; (3) within nonclassical space-time, whose structure is derived from nonclassical generative elements.

In conclusion, I would like to return to the hypothesis that space and time are macroscopic in nature. This hypothesis will be weakened by any attempt to assign it a philosophical status, as it is particular and purely physical and does not contradict the philosophical thesis that space and time are universal. This is how it is understood by most Marxist philosophers.[64] But regarding the macroscopic nature of space and time within the framework of physical problems may serve to reinforce the following hypothesis: The notion that space and time are macroscopic does not imply that the macroworld has a corresponding spatiotemporal nature. This is possessed only by a certain subset of the macroworld that we study and include in our practice within the framework of classical spatiotemporal concepts. Lately, however, we have also begun to penetrate the nonclassical macroworld (macroscopic quantum phenomena, dissipative systems, biological organization, etc.). The macroworld is not exhausted by the world of classical physics; nor are the space and time of the macroworld exhausted by the metric properties of the Euclidean space physicalized in classical mechanics. Lenin's analysis of matter illustrates a possible solution of such philosophical problems. He correctly emphasized that

> the *sole* "property" of matter with whose recognition philosophical materialism is bound up is the property of *being an objective reality* existing outside our mind.[65]

This is precisely the kind of approach needed to study the status of space and time as forms of *existence* of matter. The statement that space and time as forms of existence of matter must be embodied in the Euclidean space and time of classical physics is analogous to the identification of matter and substance. It can be said that the only "property" of space and time that philosophical materialism recognizes is that they are a form of existence of matter, which, in turn and in the same context, has the single property of being objective reality. This, of course, does not imply that matter, space, and time possess only these properties[66] but that these are the properties with the recognition of which philosophical materialism is connected. Other properties (of a different level of concreteness) can be and are refined and changed through advances in scientific knowledge and dialectical materialism. Moreover, matter figures concretely in different physical theories as substance, fields, particles, etc., and the form of existence of matter is manifested as the various empirical and theoretical structures embodied in Euclidean, Riemannian, Hilbert, and other spaces, with allowance for their metric, topological, set-theoretical, and other properties. The development of the theoretical structures of physical theories corresponds to the evolution of their empirical structures. *Remaining in the solidified macroworld would not give us deeper or fuller knowledge of the micro- and megaworlds.* Let us not forget that the division into micro-, mega-, and macroworlds is fairly arbitrary and is historically changeable—lately science has arrived at the notion of the dialectical "bootstrap nature" of an integral world in which the elementary particle may prove to be the semiclosed universe, and mankind's distinctive features may determine which physical constants are selected to describe the structure of the universe.

Space and Time in the Axiomatic Approach to Physical Theory

In the preceding section I considered the status of space and time and directions in their development in modern physics. Lately, however, physical theories themselves have become the object of intense logicomathematical scrutiny, and attempts have been made to reconstruct them in an axiomatic form that, although attractive, is actually a kind of program. In a certain sense it can be stated that axiomatized physical theories are the tomorrow of physics; as for today, we have thus far not progressed beyond axiomatizing the mechanics of the mass point. This in turn accounts for the relevance of studying the status of space and time in the axiomatic approach to physical theory. Such an investigation may open up important aspects in the future development of spatial and temporal conceptions and of physical axiomatics itself.

Presenting physical theory in the form of an axiomatic deductive theory is indeed alluring: All knowledge is implicated in compact structures and is explicated by means of theorems. However, it is far from always possible to do so in formalized language, such as the first-order predicate calculus. Thus far we have not gone beyond simple physical theories; these theories have been successful, for example, for the classical mechanics of the mass point. Reconstructing logically complex physical theories has proved considerably more difficult and less fruitful. The structure of a deductive theory is a logical structure in a pure form. L. B. Bazhenov emphasizes that, when we inquire as to the structure of a physical theory, we are speaking of the structure of the deductive theory, which can be "illumined" in a given physical theory or in its logical reconstruction, when it is, if you will, "deductivized."[67]

Such "deductivization" is in fact a kind of program. It is difficult not only in the case of mature physical theories but even in applied mathematics. V. V. Nalimov observes in this connection that, even if many of the formulations in the logical structure of applied mathematics are clothed in axioms and even if the deductions from them are put into theorems, we will still not be dealing with integral structures (which are characteristic of applied mathematics).[68] In certain cases they will have been exchanged for a brilliant mosaic of criteria, and the whole question of consistency, which is so important in the structure of pure mathematics, becomes meaningless for such a variegated collection of axioms. In other cases statements based on vague intuitive considerations are merely being written in mathematical language. The chain of syllogisms that is at least an external feature of constructions in pure mathematics has completely disappeared.

This pseudoaxiomatization is of no value to the study of the logical structure of theory. Actually, at an initial stage many physical (mathematicized) theories were just such pseudo-axiomatic systems. These were subsequently subjected to logicomathematical reconstruction, which produced rigorously formulated physical theories with consistent foundations, explicated concepts, etc. In a certain sense, an example of such development would be the transition from the geometrism and imprecise concepts of Newton's mechanics to Lagrange's analytical mechanics. Further logical reconstructions or "deductivizations" are then possible. Such, for instance, is the formalized axiomatics of classical particle mechanics.[69]

Pseudoaxiomatization is of interest to the historian of science, and "deductivization" is dear to the heart of the logician. The physicist and methodologist, on the other hand, are more likely to be interested in the logicomathematical reconstruction of physical theories, including its manner of realization, the discovery of fundamental conservation laws, and principles of invariance and symmetries. In such a form physical theories are closed systems of concepts connected by axioms, each of which describes a certain area of natural phenomena. M. E. Omel'-yanovskiy distinguishes such closed systems in modern physics as classical mechanics, the theory of electromagnetism, and the theory of relativity, all of which he regards as axiomatic (physical) systems.[70]

Thus understood, physical axiomatics is generally not constructed on rigorous logical deduction but is based on physical and mathematical implication, the use of model concepts, and so on. Such approximate axiomatizations have their good points. At any rate, if the axiomatization of physical theory is not only regarded as the logical problem of ascertaining the relations between the sentences of the theory but also treated in a wider context, then as C. G. Hempel, for instance, notes:

Whatever philosophical illumination may be obtainable by presenting a theory in axiomatized form will come only from axiomatization of some particular and appropriate kind rather than just any axiomatization or even a formally especially economic and elegant one.[71]

Physical axiomatics is treated variously in the contemporary literature on logic and scientific methodology. One such work deserving special mention is Mario Bunge's interesting *Philosophy of Physics,* which has been justly regarded as "a kind of eulogy to axiomatics in physics."[72] Containing a general description of axiomatics and the present state of axiomatization techniques in physics, Bunge consistently pursues the idea that there is little hope of ordering and substantiating physical concepts outside an axiomatic system. Bunge's approach is quite fruitful for the study of the status of space and time in axiomatized physics. Physical axiomatics does not transform physical theory into a calculus; instead, from the outset it contains an informal component and a heuristic load, because it "makes a firm semantical commitment at the axiom level and carries it through all the way down to the theorems."[73]

Bunge distinguishes certain ideas that do not belong to physics but are necessary to any physical theory: (1) formal premises (logic and mathematics), (2) philosophical premises (semantics and metaphysics), and (3) protophysics, which includes a general theory of space and time. As Bunge notes, however, notions of space and time are often ranked as primary undefinable concepts. In his opinion, there are three weighty reasons for regarding protophysical conceptions of space and time as primary in a particular physical theory. First, there is uniformity, because all "variables" (actual sets and functions) of the theory that are assigned a physical meaning are listed and thus regimented and watched over. Second, different theories may require different concepts of space and time (theoretically, it is not ruled out that some of them may not need any such concepts at all). And finally, protophysical notions are often "shrouded by a fog that can be lifted only if we scrutinize them often enough."[74]

Taken together, such undefinable concepts of physical theory constitute its primary base. As regards the definition of their mathematical status, physical meaning, and interrelation, these functions are performed by axioms, which are subdivided accordingly into formal (mathematical), semantic, and physical axioms.

Although this scheme of axiomatization is difficult to realize, it is well ordered and consistent. Many points concerning the status of space and time, however, remain unclear. If we introduce the notions of space and time as undefinable primary concepts, how do we define various conceptions of space and time? To all appearances, spatial and temporal concepts belong to the ontology of physical theory. Such an ontology can be construed as a system of constructs of the theory, the ordered set of which is regarded as a model of the real world.[75] In this sense, for instance, the ontology of Newtonian mechanics includes absolute space and time.[76] The ontology of a physical theory finally crystallizes when the theory is completed and defined with the framework of a corresponding picture of the world. In the logical reconstruction of a physical theory, of course, we can project the final product, which is represented by the ontology, onto its foundation, or protophysics. In other words, we replace the initial fuzzy ontology of a developing theory with the precise ontology of the completed theory. Properly, this is what Newton did when he gave absolute space and time the rank of axioms, of laws of motion. This realized model may have served as a prototype for Bunge, but there is much in his notions that remains unclear. In particular, there is the proposition that it is possible to take account of a given conception of space and time when transferring protophysical spatial and temporal concepts (where they are defined within the framework of the general theory) to the level of the undefinable primary concepts of a specific physical theory.

We must have a clear idea of what space and time Bunge is including in protophysics. Are they perhaps the structure of the space and time that operates in the same physical theory that we are attempting to reconstruct? But space and time are important elements in the structure of the theory and are defined by the appropriate axioms. Here it can be observed that axiomatically defined space and time still do not bear the clear imprint of the theory; they acquire this feature not in protophysics but, for example, through dynamics. Perhaps what Bunge means is not the structure of physical or mathematical space and time but the substantial or relational notions. In that case, however, it is not altogether obvious why this protophysical space and time must be developed axiomatically within some general spatial and temporal theory.

Let us take a closer look at Bunge's concept of this general theory. In his view the goal of protophysics is to analyze spatial and temporal notions. Its task as a discipline is to verify the factual (informal) assumptions of various physical theories. Protophysics is also regarded as a division of exact metaphysics (or ontology), that is, the discipline that studies the most universal features of the world. Protophysical investigations (which Bunge feels may be based on a variety of philosophical assumptions—Kantian, empiricist, materialist, etc.) should result in a system of spatial and temporal theories that are necessarily constructed in mathematical language, preferably in an axiomatic form. Moreover, proceeding from the present state of mathematics, natural science, and philosophy, Bunge concludes that these theories should be objective, relational, and relativistic. They should be, but they do not yet exist: "The creation of the appropriate theory of space-time remains an open question."[77]

It is doubtful, however, whether such a theory (in whose definition and status one can desire more precision) could give us more information about space and time than a thorough logical and epistemological analysis of the theories themselves. Nevertheless, Bunge maintains that physics is generally unable to provide us with a complete picture of space and time, that physical theory fails when it attempts to convey to us profound knowledge about their essence. The answers to such questions, he says, should be sought in the foundations of physics, which he also calls metaphysics or ontology; Bunge believes that the task of specialists working in the area that is traditionally designated the philosophy of science lies in elaborating the ontology thus described, where "they will have to compete with the physicists."[78]

Objections can be raised to such a standpoint.[79] Without disputing the legitimacy of elaborating the foundations of physics, I cannot agree with Bunge's view that the field is concerned with questions "which cannot be answered within the framework of science."[80] The foundations of mathematics are no less a special mathematical discipline permeated with a profound philosophical content, and the possible foundations of physics likewise constitute a special physical discipline. Physics will attain increased clarity and find novel solutions to problems of space and time in the course of its own development, not by moving beyond science into metaphysics.

These remarks should not be construed as condemning the axiomatization of physical theory. What I instead want to emphasize is that the status of space and time in the structure of physical theory is a complicated and multilayered problem. In axiomatization it is easy to observe how space and time are defined as primary concepts and in the corresponding axioms. But there is more than this to their status. Moreover, in physical theory and its various logicomathematical, geometrodynamic, and other reconstructions, not only do we encounter different aspects of the status of space and time but also—and this is important—we find that space and time themselves are not unique and that we must distinguish among the theoretical, empirical, conceptual, and other spaces and times of the theory. There are quite a few such aspects, and they are not all identically interconnected. Those spatial and temporal aspects that do not directly appear in axiomatics can be included in a protophysics, but it must not be forgotten that this is not the solution to the problem but merely a formulation of it. This is what Bunge means when he states that such a protophysical general theory of space and time does not yet exist. A rational reconstruction is capable of treating all these aspects and structures, but what it formalizes is not an individual physical theory but rather the evolution of physics itself, including the interrelationship and correspondence of different theories, the development of classical mechanics and its various logicomathematical analogs, the development of modern nonclassical theories of the micro- and megaworlds, the transition to the relativistic doctrine, and so on.

Proceeding to the level at which physical theories are contrasted can yield interesting results that have a bearing on physical axiomatics itself. Thus the idea that fundamental physical theories cannot in principle be axiomatized is significant in the nontrivial sense that an axiomatic construction of classical mechanics, for example, leads to corollaries that logically contradict the initial axioms. As M. I. Podgoretskiy and Ya. A. Smorodinskiy show, the appearance of such logical contradictions can be considered typical and a result of "inconsistencies of encounter."[81] Generally, such encounters are inadmissible in mathematical axiomatics. In physics they are both typical and provide a powerful internal stimulus to the development of physics in general and to theoretical physics in particular.

Classical mechanics is an interesting and illustrative example. This theory cannot even be formulated without the concept of the rigid body. Remove it, and we have neither notions of space and time nor standards of length and time; without these the theory of relativity becomes meaningless. Paradoxical as it may seem, however,

> the final conclusion of classical mechanics is that neither rigid bodies nor any fairly stable formations can exist. Such conclusions run fully counter to experience and are eloquent evidence of the imperfect and unclosed nature of classical mechanics.[82]

Additional external considerations must be adduced to substantiate classical mechanics. In particular, the obligatory stability of rigid bodies can be based only on quantum mechanics, and "we are therefore forced to conclude that a consistent axiomatic exposition of classical mechanics is impossible without the notions of quantum mechanics."[83] The converse is also true: No consistent construction of quantum mechanics can dispense with the concepts of classical mechanics. Thus we find ourselves in a kind of vicious circle in which there is no classical mechanics without quantum mechanics and no quantum mechanics without classical mechanics! And, although this scheme is circular, it is not closed, because it does not take into account fields, elementary particles, etc.

Observe that this interdependence of classical and quantum physics is nonsymmetric. Quantum mechanics could not have arisen and developed without classical notions. Nor can quantum mechanics be interpreted empirically outside the classical conceptual apparatus. As regards classical mechanics, because it matured before the appearance of quantum mechanics, it naturally did not need quantum-mechanical conceptions. Quantum mechanics may be required if the foundations of classical mechanics are to be consistently and profoundly analyzed. Thus, when the complex structure of the atom was discovered and it became apparent that it did not at all resemble a rigid body, the question arose as to the nature of a rigid body, and here quantum mechanics became indispensable.[84]

The unclosed character of the axiomatization of physical theory is beyond doubt. Completeness is not intrinsic even to mathematical axiomatics. To recall the lessons of Hilbert and Gödel, there is no final and perfect axiomatic system, and deeper levels of axiomatization are always theoretically possible.

In each individual instance in physics it is necessary to consider the introduction of a reasonable limitation on the "resolving power" of the axiomatics. Otherwise, as the analysis becomes more precise, we will be gradually forced to introduce all of physics, theory on theory, into the axiomatics of classical mechanics. This possibility, however, enables us to regard the unity of physics in a new light. Rather than merely a set of interconnected axiomatic systems, physics proves to be a single hierarchical axiomatic system.

Each level in the hierarchy becomes relatively complete[85] as it acquires a critical element; this element is necessary, first, to surmount the considerable difficulties that arise in the development of the theory; second, it contradicts the basic conceptions of the theory; and third, it serves as the starting point for a substantially new theory on the next level of the hierarchy. One example is Planck's quantum, which was introduced into classical physics as an ad hoc hypothesis to cope with difficulties in the study of energy distribution in the blackbody spectrum. Although it helped solve these problems, the quantum hypothesis itself contradicts the foundations of classical physics and "generates in the organism of the earlier theory a completely foreign body."[86] Planck's quantum was the element that completed classical physics and served as the starting point for the radically new theory of quantum mechanics.[87]

For the final element of any theory to be the starting point for a theory on the next level of the hierarchy, it must itself be generalized (in the process of developing the axiomatic foundations of the new level). Thus this element in classical physics is the finite-energy quantum of the oscillator, and the starting point for quantum theory is its generalized, theoretical analog—the quantum of action.[88]

Quantum mechanics belongs to a higher level of the hierarchy. It was constructed as the foundations of classical physics were reviewed in the light of the quantum hypothesis. Numerous physical and logicoepistemological inconsistencies and paradoxes had to be solved along the way, and it was necessary to refine and generalize many basic concepts that are found in both classical and quantum mechanics but have different values and meanings. The axioms of classical physics are adjusted accordingly, and new specific axioms are introduced. Many structural parameters are switched over. Thus, whereas Euclidean space and time are the theoretical structure of classical mechanics, the quantum theory was developed in infinite-dimensional Hilbert space—in the process of physical interpretation the elements of the corresponding spaces are transformed into fundamental objects of the theory. Moreover, many researchers conclude that the transition from classical to quantum physics involves a change in logic, which can appear in a certain complementarity with the topology of the corresponding space.

Descartes once described such an evolution. Deduction proceeding on intuitive foundations is completed by a return to these foundations that alters and calls them into question. This then becomes the basis (once again intuitively grasped) of further deductions, and so on and so forth. The system is a hierarchy of intuitions or, as we would put it today, a hierarchical axiomatic system.[89]

As I see it, the unity of physics can appear in the construction of a unified theory that assumes the role of a metatheoretical regulator that defines the scheme on which the hierarchical axiomatic system will develop and indicates at which step the axioms will be revised and expanded and at which stage concepts and terms will be changed and logic carried over.[90] Because the regulator defines the development of the hierarchy, it is subject to an inverse corrective influence from the empirical levels of the corresponding strata of the hierarchy, ontologies, etc. This brings to mind the state of affairs in contemporary biology, which has arrived at the notion of the evolution of evolution.

Here we have an unusual dynamic and hierarchical axiomatic system in which the set of axioms is expanded and altered step by step, where logic is switched over, and so on. Lately, however, considerable attention has been given to this expansion of our usual notions of axiomatic systems. Thus, in the deductive analysis of language, it is necessary to review our ordinary ideas of the stationary satisfiability of axioms. This leads to probability axiomatics, in which it is assumed that each axiom is satisfiable only with a certain degree of probability rather than always.[91] Another interesting instance is the case for which the deductive system and the system of interpretations are variable, or nonstationary.[92] Here as well, a metatheoretical modeling function appears that assumes a role analogous to that of the metatheoretical regulator.

This notion of the metatheoretical regulator allows us to consider the interrelationship and unity of physical theories dynamically rather than statically, as in, for example, A. L. Zel'manov's cube vertices.[93] Essentially incomplete and turned toward the future on the model of a growing hierarchical tree, physics is not simply going to wait for the researcher to fill in an empty cube vertex of the theories by synthesizing known fundamental constants. (Naturally, as new fundamental constants are discovered, one could begin to place physical theories among the vertices of an icosahedron or a dodecahedron, but this would not change the static nature of the procedure.)

The metatheoretical regulator enables us to restate the question of unified physical theory that has occupied such prominent twentieth-century physicists as Albert Einstein and Werner Heisenberg. Building a unified theory was generally presented as the construction of a universalized space. If the curvature of Riemannian space in Einstein's general theory of relativity defines gravitation, then one should attempt to construct a more universal space with parameters that account for other physical essences, such as electromagnetism and particles. A unified physical theory, however, can also be presented otherwise, as a metatheoretical regulator defining the scheme for constructing a hierarchical axiomatic theory of physics. This is not a search for a unified universalized space but a program for generating the set of geometries or different spaces (and the structures connecting them) that correspond to different physical theories.

In a certain sense, modern physics adopts such an approach in the geometrodynamic conception of superspace, in the theory of calibrated fields, and so on. The theory of calibrated fields, for example, provides a unified description of strong, electromagnetic, and weak interactions:

> Unlike other unified theories it does not aim at finding one multicomponent field function in which individual groups of components correspond to various types of interactions. What thus correspond to individual interactions are calibrated fields connected with a particular group of symmetries—a local calibrated group.[94]

Generating a set of different spaces possessing different degrees of universality (the T-structures of physical theories) does not deny space and time their basic status. This is due, first, to the spatiotemporal nature of the empirical interpretation and verification of physical theories, and second, to features in the metalanguage of scientific practice, which is ordinary unformalized language intimately connected with spatiotemporal definability.

Only in combination with an empirical and semantic interpretation can the logical reconstruction of a physical theory constitute the physical theory proper, which is hypothetico-deductive in nature. If the elements of the axiomatic system are not correlated with physical objects, the theory loses its relationship to reality. In the preceding section I considered such a correlation within the framework of operational procedures that elevate length and temporal duration to the rank of primary magnitudes. Confronting us here in their full scope are the problems that Bunge[95] assigns to protophysics: an elucidation of basic concepts of mereology through a physical interpretation of Boolean algebra and the construction of a metric physical geometry. All these problems arise in the operational "split" in the basic concepts that leads to the denial that different physical theories have a unified empirical basis and the assertion that in different theories we descend to distinct E-structures. In that view, the observational language is in each case marked by the theory, so that the hierarchical axiomatics of physics would probably be interpreted on a mosaic, multilayered empirical basis. The nonuniqueness of the empirical basis and the observational language and their dependence on the corresponding theories (both the fundamental one and the one that is constructed into the operational level) inevitably lead us to ask whether the space and time of classical physics should be regarded as the universal background against which physical theories are empirically interpreted and verified.

Bohr's well-known epistemological principle has already been cited to indicate that an affirmative answer to that question is untenable. As I see it, this principle demands that the E-structure of physical theories be macroscopic and in that sense classical (that is, not quantal, as in Bohr's day the notion "quantum" was associated with the microworld), but it does not insist on the embodiment of these structures in classical space and time. It must therefore be noted that in one of his studies Bohr not only associates the empirical level of physical theory with the conceptual scheme of classical physics but proceeds to classical logic. He emphasizes:

All well-defined experimental evidence, even if it cannot be analyzed in terms of classical physics, must be expressed in ordinary language making use of common logic.[96]

The nature of the logical space of events of physical theory has attracted increasing attention in recent years. Thus Jeffrey Bub, for example, concludes:

The significance of the transition from classical to quantum mechanics is understood as the proposal—on empirical grounds—that the logical space of events (micro-events) is non-Boolean. Just as the significance of the transition from classical to relativistic mechanics lies in the proposal that geometry can play the role of an explanatory principle in physics, that the geometry of events is not *a priori*, and that it makes sense to ask whether the world geometry is Euclidean or non-Euclidean, so the significance of the quantum revolution lies in the proposal that logic can play the role of an explanatory principle, that logic is similarly not *a priori*.[97]

Within the framework of the T- and E-structures proposed here, then, we must also take into account the logical spaces of events.

Thus space and time as structures of the empirical interpretation and verification of physical theories do not lose their basicity but develop regularly and are enriched as physics itself evolves. As for the problems connected with the E-structures of physical theories, they are organically bound up with the features of natural language based on classical logic; it is not classical space and time in the role of the E-structure that are epistemologically universal but only classical logic. (Non-Boolean logic appears in quantum mechanics only in the T-structure in the logical interpretation of Hilbert space.)

This brings us to the second aspect of the basicity of space and time. Of interest to us here are some comments by the well-known French mathematician René Thom, who asks: Why is ordinary language not axiomatized? How is it that members of the same linguistic community nevertheless share approximately the same semantic world? Ordinary language does not function any the worse even when a word or notion is expanded and realized through an informal class of equivalence. What is the source of this striking efficiency and lack of ambiguity (this is removed by the context) in ordinary language?

Meaning in ordinary language appears to be based mainly on topological criteria:

> The identity of an object or individual is expressed through the nature of the coherence of the spatio-temporal area occupied by this parameter or individual. The syntax of ordinary language, though relatively poor from the point of view of structure, describes the most frequent dynamic interactions of spatio-temporal parameters.[98]

Thom goes on to a comparative analysis of ordinary language, the language of Euclidean geometry, and formal algebraic language, concluding that Euclidean geometry is a natural intermediate (and perhaps indispensable) object between ordinary language and algebraic language, because "geometry allows a psychological outburst of syntax without sacrificing meaning, which is always preserved by spatial intuition."[99]

If Euclidean geometry is an indispensable intermediary link between ordinary language and formal algebra, however, the analogous link in physics consists in the space and time of the empirical level (that is, the E-structure), which only in the most simple case (classical physics) is Euclidean; as modern physics develops, space and time undergo regular changes and are marked by the corresponding physical theories. The space and time stamped into our language are not given once and for all but alter considerably with the evolution of language and thought. To cite the Sapir-Whorf hypothesis, widely different languages express different worlds that exist in their own specific spaces and times.[100] These facts border closely on the psychological aspects of space and time that are increasingly coming to the fore in modern physics. Significantly, the development of quantum physics is bringing scientists to the conclusion that the separation of physics and psychology is an important gap in modern science, a lacuna that can be filled by developing a more integrated view of the totality of our experience, which includes both the physical world and our personal existence.[101]

Thus, at least within the context of the present study, the circle is closing. Starting with psychology, I proceeded through mythology, religion, and philosophy to arrive at modern physics, whose conflicts proved to be intimately connected with psychology. One would like to think that this supports the relative closedness of my inquiry into the genesis of spatial and temporal notions, an investigation that has enabled us to consider the status of space and time in modern physics and the current development of this conception both in the unity of T- and E-structures of physical theories and in connection with the concept of hierarchical and dynamic physical axiomatics.

Notes

Introduction

1. V. I. Lenin, *Polnoye sobraniye sochinenii*, fifth edition (Moscow, 1969), vol. 24, 225.

2. Hermann Bondi, *Assumption and Myth in Physical Theory* (London, 1967).

3. Karl Pribram, *Languages of the Brain* (Englewood Cliffs, N.J., 1971).

4. Hannes Alfven, "Cosmology: Myth or science?" in *Cosmology, History and Theology*, Wolfgang Yourgau and Allen D. Breck, eds. (New York, 1977), 5.

5. Here I mean metamythological studies, such as Lévi-Strauss's structural typology of myths. Various metamythologies based on natural philosophy, religion, and science have been formulated at different times. Whereas religious metamythology has studied the mystical and theological components of mythology, scientific metamythology is interested in reconstructing its rational, cosmological, and other ingredients.

6. N. Ya. Vilenkin and Yu. A. Schreider, "Ponyatiya matematiki i ob"yekty nauki," *Voprosy Filosofii* 2 (1974), 120.

7. Some objects and phenomena of the macroworld also have a certain "bootstrap" character. One example is the hologram, in which information about every point of the object is distributed throughout the entire hologram and for which any tiny part of the hologram contains information about the entire object.

8. See, for example, D. Postle, *The Fabric of the Universe* (New York, 1976) and Fritjof Capra, *The Tao of Physics* (New York, 1984).

9. Albert Einstein, "The problem of space, ether and the field in physics," in *Essays in Science* (New York, 1953), 61.

10. Max Jammer, *Concepts of Space* (Cambridge, Mass., 1954); V. I. Sviderskiy, *Prostranstvo i vremya* (Moscow, 1958); A. N. Vyal'tsev, *Diskretnoye prostranstvo-vremya* (Moscow, 1965).

11. Charles H. Kahn, *Anaximander and the Origins of Greek Cosmology* (New York, 1960), 20.

12. M. A. Akhundov and L. B. Bazhenov, "Status prostranstva i vremeni v nauchnom poznanii," in *Filosofskiye problemy sovremennogo estestvoznaniya*, (Kiev, 1977), vyp. 42, 23.

13. See David Bohm, *The Special Theory of Relativity* (New York, 1905), appendix.

14. Viewing reality through different "glasses" (myth, natural philosophy, scientific theory) at various periods of history, mankind has accordingly seen different worlds.

15. B. Carter, "The coincidence of large numbers and the anthropic principle in cosmology," in *Confrontation of Cosmological Theories with Observational Data*, N. Longair, ed. (Dordrecht, 1974). The precursors of the anthropic principle are to be found in the mythological world view, in which it is allowed that the macrocosm (the universe) can be known through knowledge of the microcosm (mankind). Various forms of this idea have been adopted by religion, natural philosophy, and science.

16. I find quite justified the approach of Hüseyin Yılmaz, who takes the study of elementary phenomena of perception as the starting point for an analysis of the structure of modern physical theories. See his "Perception and the philosophy of science," *Boston Studies in the Philosophy of Science* 13 (1974), 3.

17. John Burnet, *Early Greek Philosophy* (London, 1920). F. Kh. Kessidi correctly perceives the sources of such notions in the fact that the "philosophical world view usually remains beyond the field of vision of historians of philosophy and science, which creates the impression that "in the beginning was nothing . . . and then, suddenly, as if out of Hesiod's yawning black abyss, Thales, Anaximander, and Anaximenes appeared one after the other on the horizon of European science and philosophy" (F. Kh. Kessidi, *Ot mifa k Logosu* (Moscow, 1972), 102).

18. Heinrich Ritter, *Geschichte der Philosophie alter Zeit* (Hamburg, 1836), vol. I, 357–358.

19. A. F. Losev, *Antichnaya filosofiya istorii* (Moscow, 1977), 38.

20. Marshall Clagget, *Greek Science in Antiquity* (New York and London, 1966).

21. This problem has attracted considerable attention in recent years. Gone is the time when scholars shrugged off the medieval period as simply the Dark Ages. Deserving special mention is A. Ya. Gurevich, *Kategorii srednevekovoy kul'tury* (Moscow, 1972), an interesting study of time and space as categories of medieval culture.

22. See Roger Penrose, *An Analysis of the Structure of Space-Time* (Princeton University: Palmer Physical Laboratory, undated), 3.

23. Jürgen Ehlers, "The nature and structure of space-time," in *The Physicist's Conception of Nature*, Jagdish Mehra, ed. (Dordrecht, 1973), 71–89; Andrzej Trautman, "The theory of gravitation," in *The Physicist's Conception*, 179–186.

24. Albert Einstein, "Field theories old and new," *The New York Times*, 3 February 1929 (New York: Readex Microprint Corporation Reprint, 1960).

25. Einstein, "Field theories old and new."

Chapter 1

1. S. I. Vavilov, *Glaz i solntse* (Moscow, 1976), 3.

2. G. J. Whitrow, *The Natural Philosophy of Time* (London and Edinburgh, 1961), 113–114.

3. See V. I. Lenin, *Polnoye sobraniye sochineniy*, vol. 24, 314.

4. See U. R. Eshby [W. R. Ashby], "Chto takoye razumnaya machina?" in *Kibernetika ozhidayemaya i kibernetika neozhidannaya* (Moscow, 1968), 35–40.

5. Martin Heidegger, *Sein und Zeit* (Tübingen, 1963), 65; Edmund Husserl, *Gesammelte Werke* (The Hague, 1954), vol. 6, 153.

6. *Phenomenology and Social Reality: Essays in Memory of Alfred Schutz*, Maurice Natanson, ed. (The Hague, 1970), 57–58.

7. V. I. Lenin, *Materialism and Empiro-Criticism* (Moscow, 1954), 180.

8. P. K. Anokhin, "Filosofskiy smysl problemy intellekta," *Voprosy Filosofii* 6 (1973), 95.

9. Here it must be remembered that "the organs which appear in the evolutionary process are above all those which indicate the position of the body in space" (N. A. Tikh, "K voprosu o polozhenii tela v prostranstve," *Izvestiya APN RSFSR* 86 (1956), 5).

10. V. Favorskiy, "Vremya v iskusstve," *Dekorativnoye iskusstvo SSSR* 2 (1965), 2. An interesting passage of this article describes the essence of binocular temporality:

The artist who renders space usually portrays more than he can see at any one time. He includes in his depiction a point of view, and since this point of view must be rendered truthfully, he involuntarily has to do with lateral areas and is forced to unite events taking place at different times. . . . But even if we stop the models and stop ourselves, there will still be time in our perception, since we have two eyes, i.e. binocularity.

11. B. G. Anan'yev and Ye. F. Rybalko, *Osobennosti vospriyatiya prostranstva u detey* (Moscow, 1964), 46.

12. For a more detailed treatment, see B. G. Anan'yev, *Psikhologiya chuvstvennogo poznaniya* (Moscow, 1960), 35–36.

13. N. A. Bernshteyn, *Ocherki po fiziologii dvizheniy i fiziologii aktivnosti* (Moscow, 1966), 70. Topological features are more fundamental spatial and temporal features than metric properties, and modern physics has only recently begun to master and use the topology of space and time.

14. Children who have developed in isolation from society remain on the developmental level of animals. They lack not only speech and thought but also even the specifically human form of locomotion. A man raised in isolation cannot "get up off his knees" and become a man.

15. V. I. Kochetkova, "Evolyutsiya mozga v svyazi s progressom material'noy kul'tury," in *U istokov chelovechestva* (Moscow, 1964), 238.

16. I. M. Sechenov, "German F. Gel'mgol'ts kak fiziolog," in *Izbrannyye filosofskiye i psikhologicheskiye proizvedeniya* (Moscow, 1947), 371.

17. This parallelism between ontogeny and phylogeny extends only to certain features. As Albrecht Peiper reasonably argues: "After all, on the basis of the fact that the infant is nourished exclusively on its mother's milk no one is

about to conclude that man's adult ancestors lived on milk alone" (Albrecht Peiper, *Cerebral Function in Infancy and Childhood*, Benedict Nagler and Hilde Nagler, trans. (New York, 1963)).

18. D. B. El'konin, *Detskaya psikhologiya* (Moscow, 1960), 79.

19. Anan'yev and Rybalko, *Osobennosti vospriyatiya*, 75.

20. M. D. Akhundov, "Antropogenez, patopsikhologiya i evolyutsiya predstavleniy o prostranstve i vremeni," in *Biologiya i sovremennoye nauchnoye poznaniye* (Moscow, 1975), ch. 2, 73.

21. J. von Uexküll, *Innenwelt der Tiere* (Berlin, 1920), 4; Peiper, *Cerebral Function*, 597–598.

22. See M. D. Akhundov, "Genezis predstavleniya o prostranstve i vremeni," *Filosoficheskiye Nauki* 4 (1976), 70.

23. See Anan'yev and Rybalko, *Osobennosti vospriyatiya*, 76.

24. Of interest here are visual agnosias, in which visually perceptible objects are often not perceived. The stage-by-stage recovery of sight in such cases resembles the evolution of vision in the child—from a sensation of a solid flood or haze of light to the differentiation of formless, amorphous spots, followed by the perception of contours, etc.

25. Painting in antiquity had not mastered perspective to any significant degree. Geometric perspective was not discovered until the Renaissance. Even though artists had learned its laws, they rarely obeyed its strict rules. We can therefore understand R. Gregory when he says that "it is striking how seldom accurate representation of space is found in art" (R. L. Gregory, *The Intelligent Eye* (London, 1970), 106).

26. William Stern, *Psychologie der frühen Kindheit bis zum sechsten Lebensjahre* (Leipzig, 1914), 70–71.

27. Besides analyses of the origin of spatial notions, there are a number of important studies on the microgenesis of perception that treat the order in which the spatial, dynamic, and figurative characteristics of objects are differentiated. See V. M. Velichkovskiy and N. V. Tszen, "Mikrostrukturnyy analiz vospriyatiya formy i stroboskopicheskogo divizheniya," in *Issledovaniye mekhanizmov vizual'nogo vospriyatiya* (Moscow, 1973), 37–51. There it is shown that the perception of space appears first, serving as a background and realizing the substantial conception of space in perception. Spatial perception requires 50 milliseconds from the inception of the stimulus, the perception of movement takes 100 milliseconds, the perception of form beginning to operate only after this.

28. Jean Piaget, *The Psychology of Intelligence*, Malcolm Piercy, trans. (London, 1971), 128.

29. Rudolf Arnheim, *Art and Visual Perception* (Berkeley and Los Angeles, 1974), 69. The presence of a common spatial structure does not warrant the conclusion that it is unique or constant. Spengler, for example, considers that

the whole of lyric poetry and music, the entire painting of Egypt, China and the West by hypothesis deny any strictly mathematical structure in space as felt and seen, and it is only because all modern philosophers have been destitute of the smallest understanding of painting that they have failed to note the contradiction. (Oswald Spengler, *The Decline of the West*, Charles Francis Atkinson, trans. (New York, 1937), vol. 1, 171)

30. Jean Marie Guyau, *La genèse de l'idée de temps* (Paris, 1890).

31. I. M. Sechenov, "Refleksy golovnogo mozga," in *Izbrannyye filosofskiye i psikhologicheskiye proizvedeniya*, 127.

32. This feature was typical of a long period of human evolution. As A. Ya. Gurevich notes: "Temporal relations do not begin to 'organize' events in the human consciousness until the thirteenth century. Until then, time was to a significant degree perceived spatially" ("Predstavleniye o vremeni v srednevekovoy Yevrope," in *Istoriya i psikhologiya* (Moscow, 1971), 189).

33. Peter H. Lindsay and Donald A. Norman, *Human Information Processing* (New York and London, 1973), 460.

34. A. G. Spirkin, "Proiskhozhdeniye kategorii prostranstva," *Voprosy Filosofii* 2 (1956), 92. L. G. Dragoli, "Vremennyye kategorii rechi i differentsiatsiya vremeni," in *Vospriyatiye Prostranstva i Vremeni* (Leningrad, 1969), 97–98.

35. Some observers consider that the child's perception differs so radically from that of the adult that their two worlds are, in effect, mutually incomprehensible; this explains the phenomenon of infantile amnesia, where the adult is unable to recall anything that happened to him or her at the age of two or three. See Dan Isaac Slobin, *Psycholinguistics*, second edition (Glenview, Ill., 1979), 157.

36. Michel Foucault, *The Order of Things: An Archeology of the Human Sciences* (New York, 1970), 98.

37. Piaget, *Psychology of Intelligence*, 136–137.

38. P. Fress, "Prisposobleniye cheloveka ko vremeni," *Voprosy Filosofii* 1 (1961), 54.

39. Pavlov once wrote: "By breaking down and simplifying, pathological phenomena often reveal to us things that are fused and complex in psychologically normal states" (I. P. Pavlov, *Polnoye sobraniye trudov* (Moscow, 1947), vol. 4, 365–366).

40. See, for example, B. F. Porshnev, *O nachale chelovecheskoy istorii* (Moscow, 1974), 365–366.

41. Claude Lévi-Strauss, *Anthropologie structurale* (Paris, 1958), 199.

42. It is interesting to note that patients with damage to the right hemisphere (epilepsy, for example, where the seat of the disturbance is located in that hemisphere) complain of a special kind of dream, which is in color and so realistic that they find it difficult to distinguish it from reality.

43. An entire constellation of Soviet researchers have contributed to the study of this problem, among them B. G. Anan'yev, L. Ya. Belen'kaya, N. N. Bragina, T. A. Dobrokhotova, B. V. Zeygarnik, T. A. Mering, L. G. Chlenov, and I. B. Eydinov.

44. T. A. Dobrokhotova and N. N. Bragina, *Funktsional'naya asimmetriya i psikhopatologiya ochagovykh porazheniy mozga* (Moscow, 1977), 83.

45. Although the significance of this problem has been realized for some time, it is only relatively recently, in connection with research on the weak interactions of elementary particles by Lee, Yang, and others, that modern physics has dealt with the possibility of disturbing spatial and temporal sym-

metry. V. I. Vernadskiy has noted that "dextrality/sinistrality may be regarded as an extremely sensitive indicator of the physical state of space" (V. I. Vernadskiy, "O pravizne i levizne," *Problemy Biogeokhimii* 4 (1940), 5).

46. For a more detailed treatment, see Ye. P. Kok, *Zritel'nyye agnozii* (Moscow, 1967); V. I. Korchazhinskyaya, "Varianty sindroma odnostoronney prostranstvennoy agnozii pri ochagovykh porazheniyakh mozga," *Voprosy Neyrokhirurgii* 3 (1971).

47. V. I. Korchazhinskaya and L. T. Popova, *Mozg i prostranstvennoye vospriyatiye* (Moscow, 1977), 81.

48. See T. A. Dobrokhotova and N. N. Bragina, "Prostranstvenno-vremennyye faktory v organizatsii nervno-psikhicheskoy deyatel'nosti," *Voprosy Filosofii* 5 (1975), 133–135.

49. V. L. Bianki, *Evolutsiya parnoy funktsii mozgovykh polushariy* (Leningrad, 1967), 11.

50. T. A. Dobrokhotova and N. N. Bragina, *Funktsional'naya asimmetriya*, 261.

51. T. A. Dobrokhotova and N. N. Bragina, *Prostranstvenno-vremennyye faktory*, 143.

52. Thinkers have been interested in these problems for centuries. Thus in early Greek philosophy we find the Moerae, daughters of Ananke (Necessity): Lachesis, Klotho, and Atropos. "Lachesis sings of the past, Klotho of the present, and Atropos of the future" (Plato, *The Republic*, X, 617c–d). But the important point is that each of the Moerae stands for a different type of knowledge (Xenocrates): Intellectual cognition is associated with the future, sensory knowledge with the present, and opinionative with the past. See Sextus Empiricus, *Against the Mathematicians*, VII, 149.

53. V. P. Zinchenko, V. P. Munipov, and V. M. Gordon, "Issledovaniye vizual'nogo myshleniya," *Voprosy Psikhologii* 2 (1973), 4.

54. Anan'yev and Rybalko, *Osobennosti vospriyatiya prostranstva u detey*, 15.

55. Jung writes:

According to the main law of phylogeny, psychic structure, in exactly the same way as anatomical structure, bears the marks of developmental stages completed by our primogenitors. This also occurs with the unconscious: in losses of consciousness in sleep, mental disorders, etc. psychic products surface which bear all the traces of the primitive spiritual state—not only with respect to their form, but also as regards their content, so that often we seem to be looking at fragments of mysterious ancient doctrines. (C. G. Jung, *Wirklichkeit der Seele* (Zurich, 1947), 111)

56. Franz Boas, *The Mind of Primitive Man* (New York, 1911), 122.

57. Claude Lévi-Strauss, *Structural Anthropology* (New York, 1963); see also his "Struktura mifov," *Voprosy Filosofii* 7 (1970), 164.

58. Methodologically, Lévi-Strauss's studies are thoroughly antipsychological and antihistorical. It is also interesting to note that his rich body of source materials sometimes brings him into conflict with his own positions. Thus the "mythical thought" he treats in *The Savage Mind* proves to be far from identical with the thinking of modern man.

59. A. Leroi-Gourhan, *Le geste et la parole* (Paris, 1964), vol. 1, 10.

60. See Friedrich Engels, *The Dialectics of Nature*, Clemens Dutt, trans. (New York, 1960), 284–285.

61. Karl Marx, *A Contribution to the Critique of Political Economy*, S. W. Ryazanskaya, trans. (New York, 1970), 20–21.

62. For a more detailed treatment, see A. N. Leont'yev, *Problemy razvitiya psikhiki* (Moscow, 1972), 283.

63. Gurevich, *Kategorii srednevekovoy kul'tury*, 15.

64. Gurevich, *Kategorii srednevekovoy kul'tury*, 18.

65. See Akhundov, "Genezis predstavleniya o prostranstve i vremeni."

66. Niels Bohr, "Discussion with Einstein on epistemological problems in atomic physics," in *Albert Einstein: Philosopher-Scientist*, Paul Arthur Schlipp, ed. (Evanston, Ill.: The Library of Living Philosophers, 1949), vol. 7, 209.

67. See, for example, A. F. Losev, *Antichnaya mifologiya v yeye istoricheskom razvitii* (Moscow, 1957); F. Kh. Kessidi, *Ot mifa k logosu* (Moscow, 1972); R. Wiemann, *Literaturgeschichte und Mythologie* (Berlin and Weimar, 1971); M. I. Steblin-Kamenskiy, *Mif* (Leningrad, 1976); Ye. M. Meletinskiy, *Poetika mifa* (Moscow, 1970).

68. Steblin-Kamenskiy, *Mif*, 4.

69. Steblin-Kamenskiy, *Mif*, 30.

70. Hesiod, *The Theogony*, 116. (Hesiod, *The Homeric Hymns and Homerica*, Hugh G. Evelyn-White, trans. (Cambridge, Mass.: Loeb Classical Library, 1936), 87).

71. See Ye. A. Alekseyenko, "Predstavleniye ketov o mire," in *Priroda i chelovek v religioznykh predstavleniyakh narodov Sibiri i Severa* (Leningrad, 1976), 72.

72. Lucien Lévy-Bruhl, *Primitives and the Supernatural*, Lilian A. Clare, trans. (New York, 1935), 19–36.

73. The mythological picture of the world was modeled on the human body, the human dwelling, or human society. It should be remembered, however, that ancient mankind did not draw any clear distinctions between itself, nature, and society.

74. E. Durkheim and M. Mauss, "De quelques formes primitives de classification," *Année Sociologique* 6 (1901–1902), 55.

75. Lucien Lévy-Bruhl, *How Natives Think*, Lilian A. Clare, trans. (New York, 1966), 102.

76. In certain ancient cultures (for example, in the old forms of Japanese Shintoism) the mythology of the tribal territory and significant landscape features became a mythology and deification of space itself. See N. A. Vinogradov, "Ikonograficheskiye kanony yaponskoy kosmogonicheskoy kartiny Vselennoi—mandala," in *Problema kanona v drevnem i srednevekovom iskusstve Azii i Afriki* (Moscow, 1973), 69.

77. George Thomson, *Studies in Ancient Greek Society*, (London, 1955), vol. 2, 51.

78. Meletinskiy, *Poetika mifa*, 207.

79. Ye. D. Prokof'yeva, "Staryye predstavleniya sel'kupov o mire," in *Priroda i chelovek v religioznykh predstavleniyakh narodov Sibiri i Severa*, 113.

80. V. G. Bogoraz (Tan), *Eynshteyn i religiya: Primeneniye printsipa otnositel'nosti k issledovaniyu religioznykh yavleniy* (Moscow-Petrograd, 1923), 30.

81. See *The Poetic Edda*, Henry Adam Bellows, trans. (Princeton, N.J., 1936), 1–27.

82. Steblin-Kamenskiy, *Mif*, 33.

83. Schliemann's discovery demonstrated the immense scientific value of mythology, which contains unique facts capable of shedding light on the sources of human culture. He "discovered a culture of a scope and wealth not even suspected by the later Greeks, a culture on a level higher than that of early historical Greece" (I. M. Tronskiy, *Voprosy yazykovogo razvitiya v antichnom obshchestve* (Leningrad, 1973), 66). Thus Mycenaean culture was discovered through mythology.

84. A. F. Losev, "Khaos," in *Filosofskaya entsiklopediya* (Moscow, 1970), vol. 5, 430.

85. Lévy-Bruhl, *Primitives and the Supernatural*, 48–49.

86. In different ages and different genres, the relationship between the times of "one's own" world and the world "of others" underwent significant changes. Thus, in the early Greek adventure novel, the hero and heroine fall in love instantaneously and passionately (moment t_1), but their happy marriage is delayed by tribulations: abduction of the bride, wanderings about the world, shipwrecks, war, the hero or heroine being sold into slavery, taken prisoner, etc. An enormous amount of time is spent overcoming these difficulties. Finally the pair are united and married (moment t_2). Characteristically, however, t_1 and t_2 are contiguous. The tribulations occurred in the "alien" world and are measured on its temporal scale. As M. M. Bakhtin observes: "This is a temporal hiatus between two moments of biographical time" (*Voprosy literatury i estetiki* (Moscow, 1975), 240). In the case of Voltaire's Candide, however, these times are added together.

87. Bogoraz (Tan), *Eynshteyn i religiya*, 19.

88. A. P. Elkin, "The secret life of the Australian aborigines," *Oceania* 3 (1932), 135–136.

89. Lévi-Strauss, "Struktura mifov," 154.

90. Meletinskiy, *Poetika mifov*, 173.

91. James Frazer notes that there was a certain worship of insanity or insane individuals in primitive societies that has not been sufficiently reckoned with by historians but that has exerted an enormous influence on the development or decay of the institutions of various peoples. See his *Man, God and Immortality: Thoughts on Human Progress* (London, 1927), 207. What is meant here are primarily hysteria and epilepsy (already the *Corpus Hippocraticum* denies the divine nature of epilepsy, but human insanity continued to be treated as divine intelligence even in later periods), which were the professional illnesses of the shamans, sorcerers, and other legislators of attitudinal norms in primitive society.

92. E. Leach, "Two essays concerning the symbolic representation of time," in *Rethinking Anthropology* (London, 1961), 126.

93. See Lévy-Bruhl, *How Natives Think*, 63.

94. Thus Radcliffe-Brown notes:
The Andamese, to all appearance, regard each little story as independent, and do not consciously compare one with another. They thus seem to be entirely unconscious of what are obvious contradictions to the student of the legends. (A. R. Radcliffe-Brown, *The Andaman Islanders* (Glencoe, Ill., 1948), 188)

95. R. M. Berndt and C. H. Berndt, *The World of the First Australians* (Sydney, 1964), 201–202.

96. The archaic mythology of aboriginal peoples first reached Europe through missionaries, and this is why their mythological world is split into the natural and supernatural components that are so common in and necessary to religious consciousness.

97. Yu. M. Lotman, "O mifologicheskom kode syuzhetnykh tekstov," in *Sbornik statey po vtorichnym modeliruyushchim sistemam* (Tartu, 1973), 87.

98. I. M. Tronskiy notes of the society of early antiquity that it was deeply conservative as regards ideological forms. The typical way to introduce all innovations was to present them as the restoration of a distorted, distant past (*Voprosy yazykovogo razvitiya v antichnom obshchestve*, 138).

99. B. A. Rybakov, "Kosmogoniya i mifologiya zemledel'tsev eneolita," *Sovetskaya Arkheologiya* 1 (1965), 43. In the cyclic model, time did not always flow in a single closed circle. At a certain stage in the evolution of temporal conceptions we encounter concentric circles of time. This model contains, albeit in an inchoate form, the idea of development (a succession of cycles), which is clearly expressed in a "building of bridges" between the different concentric circles. This, however, is already the spiral model.

100. T. I. Kashina, "Semantika ornamentatsii neoliticheskoy keramiki Kitaya," in *U istokov tvorchestva (Pervobytnoye iskusstvo)* (Novosibirsk, 1978), 188.

101. *Proiskhozhdeniye religii v ponimanii burzhuaznykh uchenykh* (Moscow, 1932), 172.

102. C. J. Ducasse, *A Philosophical Scrutiny of Religion* (New York, 1953), 63.

103. A. Hultkrantz, *Les religions des Indiens primitifs de l'Amerique* (Uppsala, 1963), 33.

104. The effect of different temporal orientations can be illustrated with an example taken from modern science. A biologist oriented toward the present uses a descriptivist approach, whereas an orientation toward the past leads to a mechanical description of biological phenomena, and an orientation toward the future results in teleological descriptions. See D. B. S. Kholdeyn [J. B. S. Haldane], "Vremya v biologii," *Mir Nauki* 4 (1965), 11.

105. Thus in Christianity mankind is prohibited from "looking back" (recall the sad fate of Lot's wife) and is entirely directed into the future: Ahead lies salvation, behind is disaster.

106. M. E. Mat'ye, *Drevneyegipetskiye mify* (Moscow-Leningrad, 1956), 29.

107. Ludwig Feuerbach, *Lectures on the Essence of Religion*, Ralph Manheim, trans. (New York, 1967), 118.

108. Lenin, *Polnoye sobraniye sochinenii*, vol. 24, 50.

109. Specifically, Engels noted:

A single God never could have appeared *without a single ruler*. . . . The unity of a God in control of the numerous phenomena of nature and uniting mutually hostile forces of nature is only a reflection of the single oriental despot, who apparently or really unites people with hostile, conflicting interests. (K. Marx and F. Engels, *Sochineniya*, second edition, vol. 27, 56)

110. A. P. Okladnikov, "Stanovleniye cheloveka i obshchestva," in *Problemy razvitiya v prirode i obshchveste* (Moscow-Leningrad, 1958), 143.

111. Death dictated ancient peoples' relationship not only to space but also to time. Many primitive peoples, for example, observed the taboo against uttering the names of the deceased. Such names, which generally had a specific meaning ("fire," "road," etc.) were withdrawn from use, and new ones were invented to replace them. Besides introducing an element of instability into language, this custom also disrupted the continuity of social life, rendering a chronicle of the past impossible. Thus history became impossible, and legends were limited in temporal scope to the preceding hundred years. See James G. Frazer, *The Golden Bough: A Study in Magic and Religion*, third edition (London, 1919), vol. 2, 363.

112. Paul Lafargue observes:

The spirits of the dead which inspired the savages with such dread became tutelary divinities for men of the patriarchal period. They lodged the soul of the departed in the center of their dwellings so that it would protect and take care of the affairs of the family, it offered advice to the patriarch who had succeeded it and who never made an important decision without consulting the ancestors of the tribe. (Paul Lafargue, *Le déterminisme économique de Karl Marx* (Paris, 1928), 218)

113. Y. Tuan, *Topophilia* (New Jersey, 1974); R. D. Sack, "Magic and space," *Annals of the Association of American Geographers* 66 (1976), 309–320.

114. Ludwig Feuerbach, *The Essence of Christianity*, George Eliot, trans. (New York, 1957), 179.

115. E. Shortland, *Traditions and Superstitions of New Zealanders* (London, 1856), 82; A. M. Khazanov, "Razlozheniye pervobytnoobshchinnogo stroya i vozniknoveniye klassovogo obshchestva," in *Pervobytnoye obshchestvo: Osnovnyye problemy razvitiya* (Moscow, 1975), 115.

116. M. A. Dandamayev, "Novyye dannyye o religii v Persii na rubezhe VI–V vv. do n. e.," *Vestnik Drevney Istorii* 2 (1974), 29. For a more detailed treatment of the Achaemenidians' policies, see M. A. Dandamayev and V. G. Lukonin, *Kul'tura i ekonomika drevnego Irana* (Moscow, 1980), 332–343.

117. Margarete Riemschneider, *Von Olympia bis Ninive im Zeitalter Homers* (Heidelberg, 1964), 7.

118. See M. D. Akhundov, "Prostranstvo i vremya v mifologicheskoy kartine mira," in *V. I. Lenin i nekotoryye voprosy razvitiya nauchnogo znaniya* (Baku, 1980), 172.

119. These spatial transformations have temporal analogs. Thus when the Church Fathers moved the celebration of Christ's birth from January 6 to December 25, they affected a sacral renormalization of time. The pagans (for example, in Mithraism) celebrated the birth of the sun on December 25, and Christianity "filled" this day with its own holiday in order to transfer the

religious zeal of the pagans from the sun to its god. In fact, all Christian holidays (the festival of St. George, Christ's death, All Saints' Day, etc.) occur on days observed by the pagan cults. Frazer regards this "renormalization" as a kind of compromise on the part of the Church with these rival religions (James Frazer, *The Golden Bough*, abridged edition (London, 1949), 361). It is difficult to agree with Frazer's conclusion, for the move was a question of expansion rather than of compromise.

120. Another spatial transformation deserving mention is the sacral "recentering" of a country, by which conquerors introduced the worship of new gods, established the dominance of a new religion, and so on. The great religious reformer Amenhotep IV, for example, replaced the cult of Amon with the sun cult of Aton and proclaimed himself Ikhnaton. In connection with this the new capital of Tell-Amarna was built in the desert to succeed Thebes. When the new religion suffered a military defeat, Tutankhamon's residence was moved back to the ancient center of the country at Thebes.

121. The processes described are organically connected with ethnogeny, whose stages are inadequately classified on the basis of traditional sociological and linguistic criteria. These stages can be described as a relationship of ethnic consciousness to the category of time. The evolution of ethnos here is characterized not by any particular system of temporal measurement but by their diversity. See L. N. Gumilev, "Ethnos i kategoriya vremeni," *Doklady Geograficheskogo Obshchestva SSSR* 15 (1970), 155. If the stages of ethnogeny exhibit diverse attitudes toward time, certain scholars find a typical feature of culture in the predominant conception of time. See Philipp Frank, *The Philosophy of Science* (Englewood Cliffs, N.J., 1957).

122. Multinational states were also formed in order that the member peoples might jointly reflect the aggressor's power. Efforts here extended from the military to the religious sphere and are typical of one stage on the path to polytheism.

123. G. M. Bongardt-Levin, *Drevneindiyskaya tsivilizatsiya: Filosofiya, nauka, religiya* (Moscow, 1980), 289.

124. See E. Min'kovskaya, "Egipetskaya religiya v Kushe," in *Religiya Drevnego Egipta* (Moscow, 1976), 301.

125. Talmudists stubbornly insist that the Israelites worshipped a single god from the very beginning. Such statements are incorrect and antihistorical. N. M. Nikol'skiy, for example, counts no fewer than forty-five gods (Jahveh, Elohim, Olam, etc.) in the pantheon of the precaptivity period. To these should be added the eight chief evil spirits which ranked as gods (Azazel, Belial, Mavet, and others). See N. M. Nikol'skiy, *Izbrannyye proizvedeniya po istorii religii* (Moscow, 1974), 77. If we go even further back in time, the ancient Jews (like the Canaanites and the Phoenicians) worshipped the Baalim, idols erected on elevations in the shade of oak trees. Each Baal was the lord of its locality. See M. S. Belen'skiy, *Iudaizm* (Moscow, 1974), 49.

126. Feuerbach, *Essence of Christianity*, 6.

127. Marx and Engels, *Sochineniya*, vol. 7, 370.

128. Marx and Engels, *Sochineniya*, vol. 19, 308.

Chapter 2

1. This is precisely how Plato describes Homer. See Plato, *The Republic*, X, 606e.

2. A. F. Losev, *Istoriya antichnoy estetiki: Aristotel' i pozdnyaya klassika* (Moscow, 1975), 276.

3. See M. C. Stokes, "Hesiodic and Milesian cosmogonies," *Phronesis* 18 (1963), 1–7.

4. In his interesting structural analysis of individual myths, J.-P. Vernant discusses qualitative changes in the history of ancient Greek thought. See J.-P. Vernant, *Mythe et pensée chez les Grecs* (Paris, 1971), vol. 1.

5. Brigitte Hellwig, *Raum und Zeit im homerischen Epos* (Hildesheim, 1964), 57–58.

6. Hesiod, *The Theogony*, 116–128 (*Homeric Hymns*, 87).

7. Yu. B. Molchanov, *Chetyre konseptsii vremeni v filosofii i fizike* (Moscow, 1977), 7. Note that, although the idea of Kronos—the god of time—has no precedent in the early Greek theogonies, it was widespread in the Orient in the Sidonian cosmology, Zoroastrianism, etc. As is evident from Sumerian divine geneologies, the first steps toward a notion of the god of time as the forefather of all being were taken in Mesopotamia. See M. L. West, *Early Greek Philosophy and the Orient* (Oxford, 1971), 35.

8. A. F. Losev, *Antichnaya filosofiya istorii* (Moscow, 1977), 46.

9. Thomson, *Studies in Ancient Greek Society*, vol. 2, 151–152.

10. I. D. Rozhanskiy, *Anaksagor: U istokov antichnoy nauki* (Moscow, 1972), 48.

11. F. Kh. Kessidi, *Ot mifa k logosu: Stanovleniye grecheskoy filosofii* (Moscow, 1972), 123.

12. Rozhanskiy, *Anaksagor*, 48.

13. Kessidi, *Ot mifa k logosu*, 123.

14. S. Ya. Lur'ye, *Ocherki po istorii antichnoy nauki* (Moscow-Leningrad, 1947), 21–22.

15. See John A. Wheeler, *Einstein's Vision* (Berlin-Heidelberg-New York, 1968), 51, 68.

16. Lysia, *Funeral Oration*, 19 (*Lysias*, W. R. M. Lamb, trans. (Cambridge, Mass.: Loeb Classical Library, 1960), 41).

17. It may seem that as first principles Thales and Anaximenes took concrete objects—water and air—whereas Anaximander's was the abstract *apeiron*, but this is not quite so. All three principles were to some degree abstract and supersensory. It should also be noted that a number of recent studies persuasively argue that *apeiron* is merely a specific name (in Aristotelian terminology) for the principles of the Milesians and that Anaximander regarded Kronos as the *arche*. See A. V. Lebedev, "*To apeiron*: Ne Anaksimandr, a Platon i Aristotel'," *Vestnik Drevney Istorii* 1 (1978), 44–54.

18. The notion of material substance allows a concrete investigation to link all phenomena systematically into a unified, integral subject of scientific study, and herein lies one of its most important methodological virtues. Another is

that only on the basis of this concept of substance can we understand causal connections as an internal relation of things and phenomena. See Z. M. Orudzhev, "Preryvnost' i nepreryvnost' prichinnoy svyazi," in *Sovremennyy Determinizm: Zakony Prirody* (Moscow, 1973), 190.

19. The quest for the substance of the world was a typical undertaking throughout antiquity. Besides water and air were proposed the fire of Hippasos and Heraclitus, the numbers of the Pythagoreans, the One of the Eleatics, Empedocles' elements, Anaxagoras's homoeomeries, the atoms of Leucippus and Democritus, etc.

20. Aristotle, *Metaphysics*, I, III, 983b, 5–15 (*Aristotle in Twenty-Three Volumes*, Hugh Tredennick, trans. (Cambridge, Mass.: Loeb Classical Library, 1958), vol. 17, 19).

21. It was shown in chapter 1 that archaic myths also typically posited two times. These are the pretime before the act of creation and the cyclic time of the created world. As for the beginning of Chaos, in certain ancient Greek currents, such as Orphism (whose theogony resembles Hesiod's in certain details), it was held that Chaos was generated by eternal time.

22. Strictly, the notion of infinite space does not appear until the age of Zeno and Leucippus, but its sources are present already in Milesian philosophy. Thales maintained that "space is greatest of all, because it contains everything," and Anaximander's *apeiron* is infinite. Incidentally, Democritus later used *apeiron* to designate infinite empty space.

23. Aristotle, *Physics*, III, IV 203b, 15 (*Aristotle's Physics*, Hippocrates G. Apostle, trans. (Bloomington, Ind., 1969), 48).

24. See E. N. Mikhaylova and A. N. Chanyshev, *Ioniyskaya filosofiya* (Moscow, 1966), 69.

25. For a more detailed treatment, see P. P. Gaydenko, "Tema sud'by i predstavleniye o vremeni v drevnegrecheskom mirovozzrenii," *Voprosy Filosofii* 9 (1969), 88–98.

26. Hermann Diels, *Die Fragmente der Vorsokratiker* (Zurich-Berlin, 1964), A1, 35, 36.

27. Thus in ancient Greek culture there arose the notion of three types of cognition corresponding to different levels of the world: The world accessible to the senses was the world of opinions; that available to the intellect was the world of substances; the world accessible to madness was the world of the future. These three types later developed into philosophical currents that continue to exist today as sensualism, rationalism, and intuitionism. (Husserl's intuition, for example, is intellectual but not rational.)

28. Simplicius, in *Aristotelis Physica*, 155, 23.

29. A. N. Chanyshev, *Italiyskaya filosofiya* (Moscow, 1975), 75.

30. Aristotle, *Metaphysics*, I, V, 986a (*Aristotle in Twenty-Three Volumes*, 33).

31. Stobaeus, *Eclogai* I, 21, 76.

32. Theodor Gomperz, *Greek Thinkers. A History of Ancient Philosophy*, Laurie Magnus, trans. (London, 1955), 348.

33. Gomperz, *Greek Thinkers*, 348.

34. A. D. Aleksandrov, "A general view of mathematics," in *Mathematics: Its Content, Methods, and Meaning,* A. D. Aleksandrov, A. N. Kolmogorov, and M. A. Lavrent'yev, eds. (Cambridge, 1963), vol. 1, 26.

35. Clemens Alexandrinus, *Stromateis,* V, 105. It may seem surprising that Heraclitus should assert in this thesis that no one person created the Cosmos. According to Heraclitus, however, there are two Cosmos: the objective, unified Cosmos that is one and the same for everything that exists (which is what is meant) and the subjective Cosmos of dreams and hallucinations. In other words, there are worlds accessible to the intellect and to insanity, and each exists in its own space and time.

36. See Thomson, *Studies in Ancient Greek Society,* vol. 2, 277.

37. Aetius, *De placitis philosophorum,* I, 28, 1.

38. Molchanov, *Chetyre kontseptsii vremeni v filosofii i fizike,* 9.

39. The path from Chaos to Cosmos is as follows: At the first stage the world exists in space and time (the substantial conception), and at the second stage space and time are found in the world (the attributive notion). The infinity of the universe, therefore, of necessity implies the infinity of Cosmos. Cosmos is a part of the universe (in some systems it is the entire universe), into which order, rhythm, and metrics are introduced.

40. Plutarch, *De Ei Delphico,* 8, 388E.

41. Kessidi, *Ot mifa k logosu,* 176.

42. Plutarch, *De Ei Delphico,* 18, 322B.

43. *Hegel's Lectures on the History of Philosophy* (London-New York, 1963), vol. 1, 283.

44. C. Ramnoux, *Vocabulaire et structures de pensée archaïque chez Heraclite* (Paris, 1959), ix.

45. Plotinus, *The Enneades,* IV, 8, 1 (third edition, Stephen MacKenna, trans. (London, 1956), 357).

46. Hippolytus, *Refutatio Omnium Haeresium,* IX, 9.

47. Philip Wheelwright, *Heraclitus* (Princeton, N.J., 1959), 19.

48. Clemens, *Stromateis,* V, 110.

49. Sextus Empiricus, *Against the Physicists,* vol. I, 144 (*Sextus Empiricus in Three Volumes,* K. G. Bury, trans. (Cambridge, Mass., 1936), vol. 3, 77).

50. Simplicius, *Physica,* 22, 9.

51. Space and time are structures of relations only of a partially ordered plural world. They do not exist in Chaos, nor are they found in the One of the Eleatics. What relations can there be in absolute Chaos or absolutely uniform emptiness? These two states of the world are structureless. Properly, the thesis on the absence of relational space and time in the One is as trivial as the proposition that it is impossible for one person in solitude to start a family.

52. Hermann Diels, *Doxographi Graeci* (Berlin, 1879), 480. It is interesting to note that Plato's *Parmenides* discusses the transition from rest to motion and shows that, if motion and rest are in time, then the passage between them, being neither rest nor motion, is outside of time (Plato, *Parmenides,* 156d, e).

53. Sextus Empiricus, *Against the Mathematicians*, vol. 8, 110–111 (*Sextus Empiricus in Three Volumes*, vol. 2, 57).

54. Sextus Empiricus, *Against the Mathematicians*, vol. 8, 111 (*Sextus Empiricus in Three Volumes*, vol. 2, 57).

55. True, in Anaxagoras's system there was no void either, but this did not entail eliminating motion from the world. Instead, it is realized in the form of cosmic vortices. Anaxagoras's Being is also motionless in the beginning, but Nous subsequently introduces motion into it.

56. *Plato and Parmenides. Parmenides' Way of Truth and Plato's Parmenides*, Francis MacDonald Cornford, trans. (London, 1950), 35–36.

57. Molchanov, *Chetyre kontseptsii vremeni v filosofii i Fizike*, 10.

58. Simplicius, *Physica*, 139, 5.

59. Aristotle, *Physics*, VI, 9, 239b, 15 (*Aristotle Physics*, 122). V. Ya. Komarova has reinforced this aporia in a way that discredits Achilles' athletic prowess even more: "if the tortoise turns toward Achilles and moves even at the same speed as he, they will still never meet" (*Stanovleniye filosofskogo materializma v Drevney Gretsii* (Leningrad, 1975), 91).

60. Whitrow, *The Natural Philosophy of Time*, 151. One feature of the characters in Zeno's paradoxes deserves mention: They are in fact likened to mathematical objects and participate in purely mathematical infinite division. Incidentally, it was quite clear to Aristotle that mathematical objects are motionless and that motion is proper only to physical objects. Platonism also developed similar notions.

61. Aristotle, *Physics*, VI, 239b, 35 (*Aristotle's Physics*, 123).

62. Whitrow, *The Natural Philosophy of Time*, 153.

63. Aristotle, *Metaphysics*, I, IV, 985b, 5 (*Aristotle in Twenty-Three Volumes*, XVII, 31).

64. Aristotle, *Metaphysics*, XII, VI, 107b, 30–35 (*Aristotle in Twenty-Three Volumes*, X–XIV, 143).

65. See S. Ya. Lur'ye, *Teoriya beskonechno malykh u drevnikh atomistov* (Moscow-Leningrad, 1935), and *Demokrit* (Leningrad, 1970). See also M. D. Akhundov, "K voprosu o matematicheskom atomizme Demokrita," *Filosofiya Nauki* 4 (1970).

66. N. N. Zalesskiy, *Ocherki istorii antichnoy filosofii* (Leningrad, 1975), part 1, 36.

67. B. G. Kuznetsov, *Istoriya filosofii dlya fizikov i matematikov* (Moscow, 1974),

68. V. Ya. Komarova, *Stanovleniye filosofskogo materializma*, 121.

69. Epicurus, *To Herodotus*, 59 (*Epicurus. The Extant Remains*, Cyril Bailey, trans. (Oxford, 1926), 35).

70. A. O. Makovel'skiy, *Drevnegrecheskiye atomisty* (Baku, 1946), 59.

71. For a critical analysis of arguments against *amera*, see M. D. Akhundov, *Problema preryvnosti i nepreryvnosti prostranstva i vremeni* (Moscow, 1974), 26–33.

72. See V. P. Zubov, *Razvitiye atomisticheskikh predstavleniy do nachala XIX veka* (Moscow, 1965), 16.

73. Sextus Empiricus, *Against the Mathematicians*, VII, 139 (*Sextus Empiricus in Three Volumes*, vol. 2, 77).

74. Losev, *Istoriya antichnoy estetiki*, 431.

75. S. Ya. Lur'ye, *Ocherki po istorii antichnoy nauki* (Moscow-Leningrad, 1947), 169.

76. Erich Frank, *Plato und die sogennanten Pythagoreer* (Halle, 1923), 53.

77. Makovel'skiy, *Drevnegrecheskiye atomisty*, 91.

78. See F. I. Shcherbatskiy, *Teoriya poznaniya i logika po ucheniyu pozdneyshikh buddistov* (St. Petersburg, 1909), ch. 2, 103.

79. Numerous scholars persist in writing about the geometric properties of empty space. See, for example, V. I. Sviderskiy, *Prostranstvo i vremya* (Moscow, 1958), 8.

80. Aristotle, *Physics*, IV, 8, 215a (*Aristotle's Physics*, 73).

81. Lur'ye, *Teoriya beskonechno malykh*, 60–61.

82. Komarova, *Stanovleniye filosofskogo materializma*, 116.

83. See A. N. Vyal'tsev, *Diskretnoye prostranstvo-vremia* (Moscow, 1965).

84. Epicurus, *To Herodutus*, 61 (*Epicurus, The Extant Remains*, 37).

85. Aristotle, *Physics*, IV, 8, 216a, 20 (*Aristotle's Physics*, 75).

86. Sextus Empiricus, *Outlines of Pyrrhonism* (*Sextus Empiricus in Three Volumes*, vol. I, 381).

87. Aristotle, *Physics*, VI, 1, 232a, 5 (*Aristotle's Physics*, 106).

88. Alexander Aphrodisiensis, in Aristotle's *De Sensu*, 56, 13.

89. *Scholia Coisliniana*, in Aristotle's *De Caelo*, I, 1, 268a, 1 (Scholia cod Coisl. 166), 469e, 14.

90. Plato, *The Timaeus*, 55c (R. G. Bury, trans. (London-Cambridge, Mass.: The Loeb Classical Library, 1952), 135).

91. Plato, *Timaeus*, 58d (Bury's translation, 145).

92. See, for example, V. Geizenburg [W. Heisenberg], "Razvitiye ponyatiy v fizike XX stoletiya," *Voprosy Filosofii* 1 (1975), 88.

93. A. F. Losev, *Antichnyy kosmos i sovremennaya nauka* (Moscow, 1927), 18.

94. See V. F. Asmus, *Platon* (Moscow, 1969), 133.

95. See Aristotle, *Physics*, IV, 2, 209b, 10–15 (*Aristotle's Physics*, 61).

96. Generally, Plato's theory of knowledge is more complicated and is not divided into two levels (reason and sensation). We must also take into account his doctrine of Eros, in which he developed the notion of a special mystical trance, which the philosopher enters in order to rise to the world of ideas. Plato wavers on the question of how the Idea is to be comprehended—at times it is accessible to the intellect, at times to madness (mania, frenzy, ecstasy, etc.). See, for example, Plato, *The Symposium*, 203a.

97. Plato, *Timaeus*, 37d–38a (Bury's translation, 77).

98. Note here that the terms "time" and "eternity" in Plato's usage have a certain polysemy, and he did not always use them in the same sense in all his

works. See E. Maula, *On the Semantics of Time in Plato's Timaeus* (Abo, 1970), 32.

99. Aristotle, *Physics*, IV, 10, 218a, 5–10 (*Aristotle's Physics*, 78–79).

100. Aristotle, *Physics*, IV, 11, 220a, 5 (*Aristotle's Physics*, 82).

101. Aristotle, *Physics*, VI, 10, 218b (*Aristotle's Physics*, 124).

102. Aristotle, *Physics*, IV, 10, 218b (*Aristotle's Physics*, 79).

103. Aristotle, *Metaphysics*, X, 1, 1053a, 10 (*Aristotle in Twenty-Three Volumes*, X–XIV, 9). It is interesting to note that the sun was the only metricizer of time already among the ancient Egyptians, who depict it as follows in their sunrise hymn: The sun moves uniformly, continually, with maximum speed (the sun is a runner no one can overtake) and determines the basic rhythm of things (day/night, life/death, death/resurrection). See I. G. Lifshits, "Vremya-prostranstvo v yegipetskoy iyeroglifike," in *Akademiya nauk SSSR akademiku N. Ya. Marru* (Moscow, 1935), 231–234.

104. Aristotle, *Physics*, IV, 10, 218b, 15 (*Aristotle's Physics*, 80).

105. Molchanov, *Chetyre kontseptsii vremeni v filosofii i fizike*, 17.

106. Aristotle, *Physics*, IV, 12, 221b, 5–10 (*Aristotle's Physics*, 84).

107. Aristotle, *Metaphysics*, XII, 7, 1073a, 5–10 (*Aristotle in Twenty-Three Volumes*, X–XIV, 153).

108. Aristotle, *Physics*, IV, 11, 219b, 1–5 (*Aristotle's Physics*, 80).

109. See, for example, *Filosofiya yestestvoznaniya* (Moscow, 1966), part 1, 138.

110. D. V. Dzhokhadze, *Dialektika Aristotelya* (Moscow, 1971), 193.

111. Aristotle, *Physics*, IV, 1, 208b, 5 (*Aristotle's Physics*, 59).

112. Aristotle, *Physics*, IV, 1, 209a (*Aristotle's Physics*, 60).

113. Aristotle, *Physics*, IV, 1, 208b, 25–30 (*Aristotle's Physics*, 60).

114. Aristotle, *Physics*, IV, 1, 208b, 10–20 (*Aristotle's Physics*, 59–60).

115. "He, Democritus, holds that 'atoms' . . . move in infinite empty space, in which there is no up, no down, no middle, no end, no limit" (Cicero, *De Finibus Honorum et Malorum*, I, 6, 17).

116. B. G. Kuznetsov, *Puti fizicheskoy mysli* (Moscow, 1968), 67.

117. See I. Lakatos, *Proofs and Refutations. The Logic of Mathematical Discovery*, J. Worrall and E. G. Zahar, eds. (Cambridge, 1976).

118. Sextus Empiricus, *Against the Mathematicians*, X, 2–3 (*Sextus Empiricus in Three Volumes*, vol. 3, 211).

119. Plotinus, *Enneades*, III, 7, 8–10 (MacKenna's translation, 228–233).

120. Plotinus, *Enneades*, III, 7, 11 (Mackenna's translation, 234).

121. Molchanov, *Chetyre kontseptsii vremeni v filosofii i fizike*, 29.

122. See A. F. Losev, "Platon," *Filosofskaya entsiklopediya* (Moscow, 1967), vol. 4, 276.

123. K. Marx and F. Engels, *Sochineniya*, vol. 21, 495.

124. Bertrand Russell, *A History of Western Philosophy* (London, 1984), 420.

125. For details, see *Tvoreniya origena* (Kazan', 1899); Burton Z. Cooper, *The*

Idea of God: A Whiteheadian Critique of St. Thomas Aquinas' Concept of God (The Hague, 1974).

126. Nikolay Kuzanskiy [Nicholas of Cusa], *Izbrannyye filosoficheskiye sochineniya* (Moscow, 1937), 97–100.

127. Etienne Gilson, *La philosophie du moyen age* (Paris, 1944), 460.

128. The anti-Aristotelian elements in medieval philosophy were imposed from without and were developed by philosophers adhering to Aristotelian positions. Other medieval currents consistently objected to Aristotle from an atomistic standpoint. Among these were the terminalist systems that arose in the first half of the fourteenth century and were represented by such prominent thinkers as Walter Catton, Gerard Odonis, and Nicolas of Autrecourt. They opposed the notion of discrete space and time to Aristotle's continuous concept. See V. P. Zubov, *Razvitiye atomisticheskikh predstavleniy do nachala XIX veka* (Moscow, 1965), 102–105.

129. Alexandre Koyré, "Le vide et l'espace infini au XIV siècle," *Archives d'histoire doctrinale et littéraire du moyen age* (Paris, 1949), vol. XVII, année 24, 91.

130. Gurevich, *Kategorii srednevekovoy kul'tury*, 53.

131. St. Augustine, *Confessions*, vol. II (bk. XI, ch. XIV–XVIII) William Watts, trans. (Cambridge, Mass., and London, 1946), 247–249). St. Augustine and pseudo-Dionysius formulated two views of the world—the Latin and the early Byzantine, respectively—which exerted considerable influence on medieval thinkers. First, there is the "world as history," in which history (naturally, "sacred history") is regarded as an acute clash of opposites and the path leading from one dialectical step to another; second, there is the "world as cosmos," which is viewed as a structure, a lawful cosubordination of sensory and supersensory, a hierarchy inevitably present in timeless eternity. (See S. S. Averintsev, "Poryadok kosmosa i poryadok istorii v mirovozrenii rannego srednevekov'ya," in *Antichnost' i Vizantiya* (Moscow, 1977), 277.

132. St. Augustine, *Confessions*, vol. II (bk. XI, ch. XIV), 239.

133. We seem to encounter a temporal inversion in the time of the early Russian epics and chronicles: "The past was somewhere on up ahead, at the beginning of events, whose series had no relation to the subject perceiving it. Events 'to the rear' were those of the present or future" (D. S. Likhachev, *Poetika drevnerusskoy literatury* (Moscow, 1979), 254). Observe here, however, that the unusual arrangement of the past and future (the past in front and the future behind) derives from the fact that we are looking from outside time at a temporal segment *AB*, whose front *A* is the past and rear *B* corresponds to the present or future. If we take an event in the segment *AB*, then *A* will be behind and *B* ahead, because the chain of events (*syuzhet*) develops and time flows from *A* to *B*. All the paradoxical expression "the past is ahead" actually means is the trivial fact that the past comes "first."

134. Molchanov, *Chetyre kontseptsii vremeni v filosofii i fizike*, 32.

135. See St. Augustine, *Confessions*, vol. II (bk. XI, ch. XX), 256.

136.

Let no man say unto me hereafter, that the motions of the celestial bodies be the times, because that when at the prayer of a certain man [Joshua] the sun had stood still, till he could achieve his victorious battle, the sun stood still indeed but the time went on; for in a certain space of time of his own (enough to serve his turn) was that battle strucken and gotten. I perceive time therefore to be a certain stretching. (St. Augustine, *Confessions*, vol. II (bk. XI, ch. XXIII), 263)

137. See Yu. Borgash, *Foma Akvinskiy* (Moscow, 1966), 150.

138. Molchanov, *Chetyre kontseptsii vremeni v filosofii i fizike*, 35.

139. Engels, *Dialectics of Nature*, 184.

140. Perspective is the theory of how the volumetric and spatial properties of objects are reflected on a plane. The notion of perspective changed during the transition from the Middle Ages to the Renaissance. See L. F. Zhegin, *Yazyk zhivopisnogo proizvedeniya* (Moscow, 1970); B. V. Raushenbakh, *Prostranstvennyye postroyeniya v drevnerusskoy zhivopisi* (Moscow, 1975), 52. Medieval painting was organized on inverse perspective and moreover in a sacredly heterogeneous space in which the sacred significance of the figures determined their dimensions. Renaissance painting, on the other hand, discovered linear perspective and portrayed the real world in a homogeneous space.

141. A. Kh. Gorfunkel', *Gumanizm i naturfilosofiya ital'yanskogo Vozrozhdeniya* (Moscow, 1977), 187.

142. Bernardino Telesio, *De rerum natura juxta propria principia* (Cosenza, 1965), bk. 1, 218.

143. Francis Bacon, *Works*, James Spedding, Robert Leslie Ellis, and Douglas Devon, eds. (London, 1861), vol. 5, 496–497.

144. Francesco Patrizi, *Nova de universis philosophia* (Ferrara, 1591), 65.

145. Giordano Bruno, *Opera latina conscripta* (Naples-Firenze, 1891), bk. 3, 3.

146. Nicholas Copernicus, *Complete Works*, Edward Rosen, trans. (Warsaw-Krakow, 1978), vol. 2, 170.

147. Dzh. Bruno, *Dialogi* ("O beskonechnosti, Vselennoi i mirakh") (Moscow, 1949), 311–312.

148. Another reason Bruno was close to the conception of empty space had to do with his adherence to atomism, of which he developed diverse aspects such as the monad, the atom, and the point. See S. Ya. Lur'ye, "Teoriya nedelimykh elementov prostranstva u Dzhordano Bruno," in *Voprosy istorii fiziko-matematicheskikh nauk* (Moscow, 1963), 112.

149. Copernicus, *Works*, 170.

150. Yu. B. Molchanov, *Chetyre kontseptsii vremeni v filosofii i fizike*, 39.

151. P. Gassendi, *Sochineniya* (Moscow, 1968), vol. 2, 641.

152. Bruno, *Dialogi*, 351.

153. L. A. Foley, *Cosmology Philosophical and Scientific* (Milwaukee, 1962), 133.

154. B. E. Bykhovskiy and G. I. Naan, "Neotomistskaya naturfilosofiya i nauka," in *Nauka o neorganicheskoy prirode i religiya* (Moscow, 1973), 35.

155. Johannes Kepler, "De stella nova Serpentarii," *Gesamtausgabe von Keplers Werken (Joannis Kepleri astronomi opera omnia*, C. Frisch, ed. (Frankfurt and Erlangen, 1858–1871)), vol. 2, 642; "Harmonices mundi," vol. V, 222.

156. Galileo Galilei, *Dialogue Concerning the Two Chief World Systems—Ptolemaic and Copernican*, Stillman Drake, trans. (Berkeley-Los Angeles, 1953), 223.

157. Galileo, *Dialogue*, 147.

158. See R. Feynman, R. Leighton, and M. Sands, *The Feynman Lectures on Physics* (Reading, Mass., 1966), 9–10.

159. B. G. Kuznetsov, *Razvitiye fizicheskikh idey ot Galileya do Eynshteyna v svete sovremennoy nauki* (Moscow, 1963), 82.

160. Engels, *Dialectics of Nature*, 199.

161. Moreover, Meyerson points out that the problem of atomism is implicitly contained in Descartes's fundamental principle stating that the quiddity of a body is included in spatial extension. See Emile Meyerson, *Identity and Reality*, Kate Loewenberg, trans. (New York, 1962), 248. On the correspondence between Cartesianism and atomism, see Akhundov, *Problema preryvnosti*, 66–69.

162. René Descartes, *Principles of Philosophy*, Valentin Rodger Miller and Reese P. Miller, trans. (Dordrecht, 1983), 46–47.

163. Descartes, *Principles*, 25.

164. See Benedictus de Spinoza, *The Principles of Descartes' Philosophy*, Halbert Hains Britain, trans. (LaSalle, Ill., 1961), 129–130.

165. *The Philosophical Works of Descartes*, Elizabeth S. Haldane and G. R. T. Ross, trans. (Cambridge, England, 1911–1912; reprinted New York, 1955), vol. 2, 168.

166. Whitrow, *Natural Philosophy of Time*, 155.

167. Descartes, *Principles*, 59–60.

168. Richard J. Blackwell, "Descartes' laws of motion," *Isis* 188 (1966), 221.

169. Blackwell, "Descartes' laws," 234.

170. It is significant in this regard that George Berkeley should have accused Newton's physical system of furthering atheism—not surprisingly, as the laws of the Newtonian universe did not accommodate the will or God or miracles. See L. L. Potkov, "O meste boga v yestestvennonauchnykh trudakh N'yutona," in *Nauki o neorganicheskoy prirode i religiya*, 175.

171. For a more detailed treatment of Newton's atomism, see S. I. Vavilov, *Sochineniya* (Moscow, 1956), vol. 3.

172. Albert Einstein, "The mechanics of Newton and their influence on the development of theoretical physics," *Essays in Science*, 30.

173. *Sir Isaac Newton's Mathematical Principles of Natural Philosophy and his System of the World*, Andrew Motte, trans.; revised by Florian Cajori (Berkeley, 1960), 6.

174. *The Leibniz-Clarke Correspondence* (Manchester, 1956), 37.

175. Kenneth W. Ford, *The World of Elementary Particles* (New York-Toronto-London, 1963), 102. It is not doubted that the properties of symmetry of space and time are interrelated with the laws of conservation, but views differ as to their subordination. See R. A. Aronov and V. A. Ugarov, "Prostranstvo, vremya i zakony sokhraneniya," *Priroda* 10 (1978), 99–104.

176. Kuznetsov, *Razvitiye fizicheskikh idey ot Galileya do Eynschteyna*, 160.

177. Whitrow, *The Natural Philosophy of Time*, 160.

178. B. G. Kuznetsov, *Printsip otnositel'nosti v antichnoy, klassicheskoy i kvantovoy fizike* (Moscow, 1959), 33.

179. Meyerson, *Identity and Reality*, 82.

180. Losev, *Antichnyy kosmos i sovremennaya nauka*, 82. The identity of motion with infinite speed and immobility was discovered in earlier philosophical systems. Thus Bruno emphasized that "to move instantaneously and not to move at all are one and the same thing" (Bruno, *Dialogi*, 325).

181. Newton, *Principia*, 6.

182. Newton, *Principia*, 6.

183. G. W. Leibniz, *Philosophical Writings*, Mary Morris, trans. (London, 1961), 103–104.

184. Leibniz, *Philosophical Writings*, 21.

185. Leibniz, *Philosophical Writings*, 22.

186. Russell, *History of Western Philosophy*, 576.

187. Whitrow, *The Natural Philosophy of Time*, 38.

188. *The Leibniz-Clarke Correspondence*, 25–26. However, Leibniz did not reject space as receptacle (the substantial notion) but objected only to the void and atoms (here he repeats Aristotle). Polemizing now with Locke rather than Clarke, he states explicitly:

If matter were composed of such parts, motion in a plenum would be impossible, just as if a room were filled with quantities of small pebbles without there being in it the least empty space. But I do not admit this.... Space should rather be conceived of as full of a matter originally fluid, susceptible of any division, and submitted indeed actually to divisions and subdivisions *ad infinitum*. (G. W. Leibniz, "Essays on the human understanding," in *Philosophical Writings*, 156)

189. *The Leibniz-Clarke Correspondence*, 77.

190. Immanuel Kant, "Universal natural history and theory of the heavens," in *Kant's Cosmogony*, W. Hastie, trans.; revised edition by Willy Ley (New York, 1968), 17.

191. Friedrich Engels, *Herr Eugen Duhring's Revolution in Science*, Emile Burns, trans. (New York, 1939), 65–66.

192. Hans Reichenbach, *The Rise of Scientific Philosophy* (New York, 1951), 42–43.

193. Immanuel Kant, "Der Gebrauch der Metaphysik, sofern sie mit Geometrie verbunden ist, in der Naturphilosophie dessen engte Probe die physische Monadologie enthalt," *Werke* (Wiesbaden, 1960), vol. 1, 533.

194. Kant, "Der Gebrauch der Metaphysik," 535.

195. Kant, "Der Gebrauch der Metaphysik," 537.

196. Immanuel Kant, *Critique of Pure Reason*, Norman Kemp Smith, trans. (London, 1958), 68.

197. Kant, *Critique*, 69.

198. Kant, *Critique*, 69.

199. Kant, *Critique*, 75.

200. Hans Reichenbach, *The Direction of Time* (Berkeley, 1956), 13.

201. Russell, *History of Western Philosophy*, 687.

202. T. E. Wilkerson, "Things, stuffs, and Kant's aesthetic," *Philosophical Review* 82 (2) (1973), 169–172.

203. Kant, *Critique*, 181.

204. Kant, *Critique*, 396.

205. Kant, *Critique*, 397.

206. Kant, *Critique*, 399.

207. See E. M. Chudinov, "Logicheskiye aspekty problemy beskonechnosti Vselennoy i relyativistskaya kosmologiya," in *Beskonechnost' i Vselennaya* (Moscow, 1969), 211.

208. *Hegel's Science of Logic*, A. V. Miller, trans. (London, 1969), 190.

209. *Hegel's Philosophy of Nature*, M. J. Petry, trans. (London, 1970), vol. 1, 231.

210. *Hegel's Philosophy of Nature*, vol. 2, 67.

Chapter 3

1. Aristotle, *Posterior Analytics*, I, 27, 35 (Jonathan Barnes, trans. (Oxford, 1975), 45).

2. S. A. Yanovskaya, *Metodologicheskiye problemy nauki* (Moscow, 1972), 174.

3. See Kant, *Critique*, 53–54.

4. Rudolf Carnap, *The Philosophical Foundations of Physics* (New York, 1966), 183.

5. For a more detailed treatment, see E. M. Chudinov, *Teoriya otnositel'nosti i filosofiya* (Moscow, 1974), 140–154.

6. F. Gonseth, *Les mathématiques et la réalité* (Paris, 1936).

7. V. S. Chernyak, "Obosnovaniye elementarnoy geometrii i praktika," in *Praktika i poznaniye* (Moscow, 1973), 282.

8. Albert Einstein, "Fundamental concepts of physics and their most recent changes," *St. Louis Post-Dispatch*, 9 December 1928, supplement.

9. See N. Burbaki [Nicolas Bourbaki], *Ocherki po istorii matematiki* (Moscow, 1963), 202.

10. G. Hamel, "Über die Grundlagen der Mechanik," *Mathematische Annalen* 66 (1909), 355.

11. See L. L. Kul'vetsas, "O popytkakh aksiomatizirovat' vremya v klassicheskoy mekhanike," *Trudy XIII Mezhdunarodnogo kongressa po istorii nauki. Sektsiya V* (Moscow, 1974), 237.

12. Joseph Louis Lagrange, *Mécanique analytique* (Paris, 1811–1815), 2 vols.

13. For a more detailed treatment, see V. P. Vizgin, *Razvitiye vzaimosvyazi printsipov invariantnosti s zakonami sokhraneniya v klassicheskoy fizike* (Moscow, 1972).

14. Ford, *The World of Elementary Particles*, 102.
15. Epicurus, *To Herodotus*, 61 (*Epicurus, The Extant Remains*, 37).
16. See Aristotle, *Physics*, IV, 8, 215a; VII, 5, 250a.
17. Eugene P. Wigner, *Symmetries and Reflections. Scientific Essays* (Cambridge, Mass., 1970), 29–31.
18. Penrose, *The Structure of Space-Time*, 3.
19. Penrose, *The Structure of Space-Time*, 4–5.
20. Newton, *Principia*, 6.
21. Besides the substantial and relational conceptions, I think it justified and convenient to bring the extensional and attributive notions into consideration. Attempts to single out new spatial and temporal concepts, however, occasionally incur criticism. Thus Yu. B. Molchanov writes:

> One could distinguish a special "accidental" conception of time, that which regards time as a property of a more fundamental essence. Doing so, however, would not contribute anything really new, since the difference between the substantial and the relational notions lies in the very fact that for the one, time is something independent and for the other it is not. (Yu. B. Molchanov, "Ponyatiye odnovremennosti i kontseptsiya vremeni v spetsial'noy teorii otnositel'nosti," in *Eynshteyn i filosofskiye problemy fiziki XX veka* (Moscow, 1979), 143)

First of all, I object to Molchanov's notion of an "accidental" conception. Space and time can be regarded as attributes of matter, but relegating them to the level of coincidental, insignificant accidents is hardly warranted. Second, it seems an oversimplification to regard the difference between the substantial and relational conceptions as merely a question of whether or not time and space are independent. Although substantiality exhausts independence, relationality does not exhaust nonindependence.

22. Albert Einstein, Foreword to *Concepts of Space and Time*, by Max Jammer (Cambridge, Mass., 1951), xiv–xv.
23. Philipp Frank, *Modern Science and Its Philosophy*, 224–225. The empiricists' exasperation over the nonobservability of absolute space suggests that they regarded that space as an empirical object because theoretical objects are nonobservable by definition. V. S. Shvyrev rightly emphasizes that not even the most perfect instrument can make a theoretical object visible (observable), because the latter exists only in the system of scientific knowledge. See V. S. Shvyrev, *Teoreticheskoye i empiricheskoye v nauchnom poznanii* (Moscow, 1978), 184.
24. Observe that Newton's absolute space appears in various hypostases: (1) theological space—the *sensorium Dei;* (2) the space of the world picture, as a vacuum or "box without sides"; (3) theoretical space—as an inertial frame of reference in which the laws of mechanics are realized.
25. Einstein, "The mechanics of Newton," 34.
26. P. Germaine, *Le Temps et la pensée physique contemporaine* (Paris, 1968).
27. See Newton, *Principia*, 6.
28. F. S. C. Northrop, *The Meeting of East and West* (New York, 1946), 76–77.
29. Adolf Grünbaum, *Philosophical Problems of Space and Time* (New York, 1963), 7.

30. Euclidean geometry is given different interpretations on the theoretical and empirical levels of classical mechanics. The geometric straight line, for example, is interpreted physically by means of inertial motion and the rigid straightedge.

31. Newton, *Principia*, 6.

32. Carnap, *The Philosophical Foundation of Physics*, 76.

33. These problems have been given competent treatment in the Soviet philosophical literature. See, for example, M. V. Popovich, *O filosofskom analize yazyka nauki* (Kiev, 1966).

34. B. Riemann, "Über die Hypothesen, welche der Geometrie zugrunde liegen," in *Das Kontinuum und andere Monographen* (New York, 1960), 19–20.

35. Einstein, "Field theories old and new."

36. See *Printsip otnositel'nosti* (Moscow, 1973).

37. L. I. Mandelshtam, *Lektsii po optike, teorii otnositel'nosti i kvantovoy mekhanike* (Moscow, 1972), 88.

38. Albert Einstein, "The general theory of relativity," in *The Meaning of Relativity*, fourth edition (Princeton, 1953), 55.

39. Grünbaum, *Philosophical Problems of Space and Time*, 420. Generally, the boundary conditions at infinity of which Grünbaum speaks are not necessary to solve the equations of the general theory of relativity. The solution of these equations can consist of analytical expressions describing the spatial structures of a constant positive curvature, in which conditions at infinity do not figure at all. This, of course, does not deny the significance of the substantial conception in the general theory of relativity. See J. L. Synge, *Relativity: The General Theory* (Amsterdam, 1960), 5–8.

40. Albert Einstein, "Address at the University of Nottingham," *Science* 71 (1930), 608–610.

41. C. Misner and J. Wheeler, "Classical physics and geometry," *Annals of Physics* 2 (1957), 525–603.

42. At the empirical level Einstein defined simultaneity operationally and was guided by the principle of observability, whereas at the theoretical level, in Minkowski's formalism, simultaneity is defined geometrically and the principle of simplicity plays the definitive role.

43. A. M. Mostepanenko, *Prostranstvo-vremya i fizicheskoye poznaniye* (Moscow, 1975), 27. Some scholars regard the relational conception of space and time as resulting directly from the extrapolation of the observability principle onto spatiotemporal notions. See, for example, E. M. Chudinov, *Teoriya otnositel'nosti i filosofiya* (Moscow, 1973), 44.

44. A. F. Zotov, *Struktura nauchnogo myshleniya* (Moscow, 1973), 175.

45. Percy W. Bridgman, *The Logic of Modern Physics* (New York, 1951), 18.

46. M. Hesse, "Duhem, Quine and New Empiricism," in *Knowledge and Necessity*, ed. by Royal Institute of Philosophy (New York, 1970), 203.

47. L. B. Bazhenov, *Stroyeniye i funktsii yestestvennonauchnoy teorii* (Moscow, 1978), 151. Ya. F. Askin has even introduced a special principle of "extrapolationability presumption." See Ya. F. Askin, "Beskonechnost' Vselennoy vo vremeni," in *Beskonechnost' i Vselennaya* (Moscow, 1969), 166.

48. Distinguishing theoretical from empirical structures or schemata in the construction of a physical theory is a relevant and frequently addressed task in contemporary logicomethodological studies. See, for example, V. S. Stepin, *Stanovleniye nauchnoy teorii* (Minsk, 1976), 80.

49. Werner Heisenberg, *Philosophical Problems of Nuclear Physics*, F. C. Hayes, trans. (London, 1952), 15.

50. Niels Bohr, "Discussion with Einstein on epistemological problems in atomic physics," in *Albert Einstein: Philosopher-Scientist*, 209.

51. C. F. von Weizsacker, "Classical and quantum descriptions," in *The Physicist's Conception of Nature*, Jasdish Mehra, ed. (Dordrecht, 1973), 636. Bohr and Pauli once accused Einstein of fanatical adherence to classical theory. Einstein's answer is typical in this context:

This accusation demands either a defense or the confession of guilt. The one or the other, however, being rendered much more difficult because it is by no means immediately clear what is meant by "classical theory." (Albert Einstein, "Remarks concerning the essays brought together in this co-operative volume," in *Albert Einstein: Philosopher-Scientist*, 675)

52. See, for example, M. D. Akhundov, *Problema preryvnosti i nepreryvnosti*, 118.

53. Penrose, *The Structure of Space-Time*, 3–5.

54. A. Z. Petrov, "Fizicheskoye prostranstvo-vremya i teoriya fizicheskikh izmereniy," in *Prostranstvo i vremya v sovremennoy fizike* (Kiev, 1968), 189.

55. David Bohm, *The Special Theory of Relativity* (New York, 1965), 187.

56. R. H. Atkin, "The cohomology of observations," in *Quantum Theory and Beyond*, Ted Bastin, ed. (Cambridge, England, 1971), 191–211.

57. See Yakir Aaronov and Aage Peterson, "Definability and measurability in quantum theory," in *Quantum Theory and Beyond*, 135–139. It is to be observed that in contrast to classical theory, quantum theory treats how physical objects may be known.

58. Attempts to substantiate the macroscopic nature of space and time, however, have both theoretical and empirical aspects, and this in fact determines how the hypothesis is applied (albeit not always in a clearly differentiated way) to the T- and E-structures of physical theory. On the one hand, it is pointed out that it is possible to construct a physical theory of the microworld that does not explicitly include space-time, whereas on the other hand, the question has to do with the unobservability of the spatiotemporal continuum in the physics of elementary particles. See, for example, T. Tati, "Concepts of space-time in physical theories," *Progress of Theoretical Physics*, suppl. 29 (1964), 4; E. T. Zimmerman, "The macroscopic nature of space-time," *American Journal of Physics* 30 (4) (1962), 101.

59. Chudinov, *Teoriya otnositel'nosti i filosofiya*, 151.

60. The theoretical description of various "slices" of the macroworld may also require going beyond the limits of classical space and time. The world of classical mechanics is macroscopic, but the macroworld is not necessarily classical.

61. Yu. B. Rumer, "Printsipy sokhraneniya i svoystva prostranstva i vremeni," in *Prostranstvo, vremya, dvizheniye* (Moscow, 1971), 118.

62. D. Finkelstein, "Space-time structure in high energy interactions," in *Coral Gables Conference on Fundamental Interactions at High Energy* (Miami, 1969), 325.

63. Roger Penrose, "Angular momentum: An approach to combinatorial space-time," in *Quantum Theory and Beyond*, 151.

64. See P. S. Dyshlevyy and V. S. Luk'yanets, "Pri kakikh usloviyakh gipoteza o makroskopicheskoy prirode prostranstva i vremeni imeyet smysl?" in *Prostranstvo i vremya v sovremennoy fizike*, 259.

65. Lenin, *Materialism and Empirocriticism*, 269.

66. The properties of space and time must be distinguished from the spatiotemporal properties of moving matter. See M. D. Akhundov and Z. M. Orudzhev, "O yedinstve preryvnosti i nepreryvnosti prostranstva i vremeni," *Voprosy Filosofii* 12 (1969), 58–59.

67. L. B. Bazhenov, "Stroyeniye i funktsii yestestvennonauchnoy teorii," in *Sintez sovremennogo nauchnogo znaniya* (Moscow, 1973), 399.

68. V. V. Nalimov, *Veroyatnostnaya model' yazyka* (Moscow, 1974), 173.

69. See, for example, J. C. McKinsey, A. C. Sugar, and P. Suppes, "The axiomatic foundation of classical particle mechanics," *Journal of Rational Mechanics and Analysis* 2 (2) (1953).

70. M. E. Omel'yanovskiy, *Dialectics in Modern Physics* (Moscow, 1979), 322.

71. C. G. Hempel, "On the 'standard conception' of scientific theories," *Minnesota Studies in the Philosophy of Science* 4 (1970), 152.

72. M. E. Omel'yanovskiy, "Vstupitel'naya stat'ya," in M. Bunge, *Filosofiya fiziki* (Moscow, 1975), 248.

73. Mario Bunge, *The Philosophy of Physics* (Dordrecht, 1973), 176.

74. Bunge, *Philosophy of Physics*, 142–143.

75. J. O. Wisdom, "Scientific theory: Empirical content, embedded ontology and Weltanschauung," *Philosophy and Phenomenological Research* 33 (1) (1972), 68.

76. I would like once again to point out that Newton's absolute space has different hypostases: (1) theological space, as the *sensorium Dei;* (2) world-picture space, as the vacuum or "box without sides"; (3) theoretical space, as the inertial frame of reference in which the laws of mechanics are realized. The symmetries of this space (and time) define the fundamental conservation laws of classical physics.

77. Mario Bunge, "Prostranstvo i vremya v sovremennoy nauke," *Voprosy Filosofii* 7 (1970), 92.

78. Bunge, "Prostranstvo i vremya v sovremennoy nauke," 91.

79. See M. D. Akhundov, L. B. Bazhenov, and E. M. Chudinov, "Kontseptsii prostranstva, vremeni, beskonechnosti i sovremennaya kosmologiya," in *Filosofskiye problemy astronomii XX veka* (Moscow, 1976), 342.

80. Bunge, "Prostranstvo i vremya v sovremennoy nauke," 91.

81. See M. I. Podgoretskiy and Ya. A. Smorodinskiy, "Ob aksiomaticheskoy strukture fizicheskikh teorii," in *Voprosy teorii poznaniya* (Moscow, 1969), 76.

82. Podgoretskiy and Smorodinskiy, "Ob aksiomaticheskoy strukture," 76.
83. Podgoretskiy and Smorodinskiy, "Ob aksiomaticheskoy strukture," 76.
84. Hermann Bondi adheres to a somewhat different point of view, considering that "it is tenable to claim that you can . . . deduce the atomic and quantal structure of matter from the fact that there are solid bodies" (Bondi, *Assumption and Myth*, 10–11).
85. Such completion, of course, is relative, because each level of the hierarchy consists of a certain number of theories and is completed along many channels. For example, the quantum of action h closes one current of classical physics, the speed of light c completes another, and so on.
86. Max Planck, "Das Verhältnis der Theorien zieinander," in his *Physikalische Abhandlungen*, vol. 3, 102–107.
87. A thorough categorical analysis of scientific theory has been undertaken by Z. M. Orudzhev, who gives the closing category a different treatment—it has the status of a fundamental, logically derivable result of the theory (the category of mechanical energy in dynamics, the category of profit in Marx's *Capital*, etc.). See Z. M. Orudzhev, "Nauchnaya teoriya kak sistema kategoriy: Razvitiye teorii," in *Dialektika i metologicheskiye problemy razvitiya nauki*, 26.
88. We can note the following stages of growth in the universality of the quantum hypothesis: (1) Planck introduces the quantum of thermal radiation; (2) Einstein shows that quanta are characteristic of all radiation; (3) Bohr uses the quantum of action as a generative element of quantum theory.
89. See Descartes, "Rules for the direction of the mind," in *Philosophical Works*, vol. I, 33–35.
90. N. Rashevskiy has arrived at a similar model on the basis of organismic sets (which are moreover subject to evolution, owing to the specialization or addition of new sets of elements). He hopes to use these to build a conceptual superstructure for the trinomial complex "physics-biology-sociology." See N. Rashevskiy, "Organizmicheskiye mnozhestva," in *Issledovaniye po obshchey teorii sistem* (Moscow, 1969), 459.
91. *Matematicheskiye voprosy kibernetiki i vychislitel'nykh tekhniki* (Yerevan, 1963), 71.
92. S. B. Krymskiy, *Nauchnoye znaniye i printsipy yego transformatsii* (Kiev, 1974), 149.
93. A. L. Zel'manov, "Mnogoobraziye material'nogo mira i problema beskonechnosti," in *Beskonechnost' i Vselennaya*, 302.
94. N. P. Konopleva, "Ponyatiye intertsii i printsip simmetrii," in *Printsip simmetrii*, (Moscow, 1978), 198.
95. Bunge, *Philosophy of Physics*, 141–142.
96. Niels Bohr, "On the notions of causality and complementarity," *Dialectica* 2 (3–4) (1948), 317.
97. Jeffrey Bub, "On the completeness of quantum mechanics," in *Contemporary Research in the Foundations and Philosophy of Quantum Theory*, C. A. Hooker, ed. (Dordrecht-Boston, 1973), 52. See also I. A. Panchenko, "Logiko-

algebraicheskiy podkhod v kvantovoy mekhanike," in *Fizicheskaya teoriya* (Moscow, 1980), 265.

98. R. Tom [René Thom], "Sovremennaya matematika—sushchestvuyet li ona?" *Matematika v Shkole* 1 (1973), 92.

99. R. Tom [René Thom], "Sovremennaya matematika," 92.

100. Linguistic relativity not only operates at the level of ethnolinguistics but also manifests itself in the language of science; thus we have Quine's notion of the determinative role of language relative to the ontology used, Carnap's conception of relativity toward linguistic frameworks, Zopf's notion of relativity vis-à-vis means of description, etc. See M. D. Akhundov and R. R. Abdullayev, "Otnositel'nost' k sistemam abstragirovaniya i dopolnitel'nost'," in *Printsip dopolnitel'nosti i materialisticheskaya dialektika* (Moscow, 1976), 77.

101. Ilya Prigogine, "Time, irreversibility and structure," in *The Physicist's Conception of Nature*, 591.

Index

Aaronov, Ya., 148
Aborigines, Australian, 42–43, 174n96
Absolute space
 Aristotelian, 86, 107, 113–114
 and atomism, 73–74
 and Christian theology, 94
 and Democritus's void, 113
 Ford on, 113
 and Kant, 119
 for Leibniz, 118
 for Newton, 96–97, 112–114, 131, 135–137
 nonobservability of, 188n23
 as *sensorium Dei*, 136
 and theory of relativity, 142
 and wave optics, 141
Absolute time
 and Aristotelian Cosmos, 98
 in classical mechanics, 104, 129
 and eternity, 47–48, 114
 as mathematical, 114
 for Newton, 112, 113, 114–116
 and relative time, 80, 99, 109
 in Renaissance, 104–106
 and St. Augustine, 98–99
"Achilles" (Zeno), 71
Action-at-a-distance, instantaneous, 80, 94, 115–116, 137. *See also* Absolute time
Agnosia
 spatial, 25–27
 visual, 169n24
Akasha, 1
Aleksandrov, A. D., 64
Alfven, Hannes, 2
Algamest (Ptolemy), 92
Amera, 74–76, 85
Amon, 52

Ananke, 60–62
Anan'yev, B. G., 15, 18, 29
Anaxagoras, 61–64, 73
Anaximander, 57, 59, 60, 177n17. *See also* Milesian school
Anaximenes, 57. *See also* Milesian school
Anderson, I. G., 45
Anokhin, P. N., 14–15
Apeiron, 59, 60, 62, 77, 177n17
Aquinas, Thomas, 59
Archetypes, 2–3, 13, 106
Aristotle. *See also* Cosmos, Aristotelian
 on *apeiron*, 60
 vs. Democritus, 84–86
 dynamics of, 77, 78, 89
 and evolution of spatiotemporal concepts, 88–90
 on fundamental principles, 59
 and geocentricity, 92–94
 spatiotemporal notions of, 72, 81–88, 99, 100 (*see also* Absolute space; Absolute time)
Arithmetic vs. geometry, 123–126
"The Arrow" (Zeno), 71, 72
Asymmetry, functional, 27
Atomism. *See also* Atoms
 and Democritus, 3, 73–74, 77–80, 85
 and Leucippus, 73–74, 79
 and Newton, 77, 111, 112
 and Pythagoreanism, 62–63
 and quarks, 74
 and Renaissance, 101
 and Sextus Empiricus, 78, 89–90
Atoms, 73–75, 116–117. *See also* Atomism
Averroës, 91

Axiomatic systems
  and Euclidean geometry, 126, 127–132
  and physical theory, 10, 152–165

Bacon, Francis, 101
Basic concepts, 132, 139–140, 144, 145, 162. *See also* Duration; Length
Bazhenov, L. B., 145, 153
Bernshteyn, N. A., 15–16
Bilateral mechanisms, human, and space, 15. *See also* Lateralization
Binocular vision, and temporal perception, 15, 168n10
Biological clock, 16–17
Black, M., 127
Blackwell, Richard J., 111
Boas, Franz, 30
Bogoraz (Tan), V. G., 5, 37, 40
Bohm, David, 5, 147, 148
Bohr, Niels, 4, 9, 18–19, 33, 146–147, 163
Bondi, Hermann, 2
Bootstrap model, 3, 166n7
Bourbaki, Nicolas, 127
Bradwardine, Thomas, 94
Bragina, N. N., 25, 26, 27
Brain, functional asymmetry of, 25–29, 41
Bridgman, P. W., 144
Bruno, Giordano, 53, 100, 102, 103, 105, 184n148
Bub, Jeffrey, 163
Buddhism, and sacred space, 52
Bunge, Mario, 10, 154–158, 162
Burnet, John, 6

Calculus, 108, 112
Calibrated fields, 162
Carnap, Rudolf, 127, 139
Carter, B., 6
Chanyshev, A. N., 62
Chaos
  as generative space, 58
  of Hesiod, 57–58, 60, 65
  in mythology, 1, 34, 55
Chew, Geoffrey, 3, 10
Children
  conceptual worlds of, 11, 22
  and Newtonian mechanics, 23
  perception in, 18–20, 170n35
  and relativity, 5, 23
  sensory systems in, 17
  and social phenomena, 16, 168n14

Christian theology
  mythological roots of, 7, 91
  space in, 53, 91–95
  time in, 95–99
Chudinov, E. M., 123, 149
Clagget, Marshall, 7
Clarke, Samuel, 112, 118
Classical mechanics. *See also* Newton, Sir Isaac
  basic concepts in, 132
  and Christianity, 99
  and quantum mechanics, 158–159
  and reconstruction, 8, 132, 133–138
  spatiotemporal conceptions in, 104, 129–130, 132, 134–138
  and special theory of relativity, 138, 143
  T-structure in, 138, 142
Classical system, definition of, 146
Cognition
  abstract, 27–29
  and Marxist-Leninist philosophy, 31–32
  sensory, 27–29, 121, 171n52
  types of, for ancient Greeks, 178n27
Conceptual splits, 145
Consciousness, human, 31–32
Conservation laws, 131
A Contribution to the Critique of Political Economy (Marx), 31
Copernicus, 100, 102–106, 130
Cosmos
  Aristotelian, 87–88, 98, 100–101, 107, 133
  Aristotelian-Ptolemaic, 92–93, 99, 102
  in Christianity, 92–94
  in Greek philosophy, 34, 56, 62, 179n39
  for Heraclitus, 64–65, 179n35
  and sociopolitical organization, 45–46
Critique of Pure Reason (Kant), 120

Dandamayev, M. A., 50
Deductivization, 153
Democritus, 73–80. *See also* Democritus's void
Democritus's void
  and Aristotle, 85–86
  and Newtonian space, 8, 113, 135–136
  and Renaissance, 101

Descartes, René, 8, 108–112, 160
Dialectical materialism, 1, 12–14, 32, 116, 124, 151
"The Dichotomy" (Zeno), 71
Dicke, P., 6
Dobrokhotova, T. A., 25, 26, 27
Duration, 109, 114, 137, 139, 162
Dynamics
 Aristotelian, 133
 characterized, 129
 evolution in, 89
 of Newton, 112

Egocentrism, and perceptual space, 20
Einstein, Albert. *See also* Relativity, special theory of
 on differential law, 112
 on Euclidean geometry, 89, 128
 and four-dimensional space-time, 142
 on Newtonian space, 136
 on scientific thought, 3
 and unified physical theory, 161
 and wave optics, 141
Eleatic school
 and change, 77
 and empty space, 73
 and Kant, 121
 and motion, 64
 and the One, 67–73
 and Pythagoreans, 62, 63–64
 and truth, 68–70
Electromagnetism, 116, 141, 161
Elements (Euclid), 8, 88, 125
Elkin, A. P., 41
El'konin, D. B., 17
Empedocles' elements, 79
Empirical structures
 and Bohr's principle, 146–147, 163
 as macroscopic, 143, 145–147, 148, 149, 163
 and physical theories, 148, 163–165
 in relativity vs. classical mechanics, 138, 145–147
Empty space. *See* Absolute space
Engels, Friedrich
 on analytic geometry, 108
 on beginnings of science, 100
 and human consciousness, 31
 on Kant's cosmology, 119
 on variable concept, 112
Enuma elish, 7
Environmental adaptation, 18

Epic historicism, 55
Epicurus, 74, 77, 89–90, 131
Erigena, Johannes, 53
Erlangen program, 128, 133
E-structures. *See* Empirical structures
Eternity
 and absolute time, 47–48, 114
 and Descartes, 109
 in medieval philosophy, 96, 99
 and Newtonian mechanics, 99, 114
 and relative time, 47
Euclid, 88, 126. *See also* Geometry, Euclidean
Euler, Leonhard, 9, 130
Evolution, human, 13–32
Existentialism, 13–14
Expectation, and time, 97–98
Extension, 111, 117
Extrapolationability, 145

Fate, 60–61
Favorskiy, V., 15
Feuerbach, Ludwig, 47, 49–50, 53
Feynman, R., 107
Field theory, and Newtonian physics, 9. *See also* Unified theory
Finkelstein, D., 150
Fire, and Cosmos, 65
Fluid multiplicity, 37
Ford, Kenneth, 113, 131
Foucault, Michel, 23
Foundations (Hilbert), 126
Frank, E., 76
Frank, Philipp, 136
Frazer, J., 173n91, 175n119
Fress, P., 23–24
Future, 21, 61, 96, 97–98

Galilei, Galileo, 106–107, 133–134
Gassendi, Pierre, 104
Geocentricity, 92–94, 102
Geometrodynamics, 1, 142
Geometry
 analytic, 108
 ancient Greek, 8, 80
 and arithmetic, 123–126
 as cosmological theory, 89
 Euclidean, 125–132, 148–149, 151, 164
 and Kant, 122
 and Kepler, 106
 mathematical vs. physical, 127
 and physics, 108, 125–129
Gilson, E., 93

Gödel, Kurt, 159
Gomperz, Theodor, 63
Gonseth, F., 127
Gorfunkel, A. Kh., 101
Gravitation, 115, 137
Greek philosophy, early, 51–62
Greek Science in Antiquity (Clagget), 7
Grünbaum, Adolf, 138, 142
Gurevich, A. Ya., 13, 32–33, 95, 167n21, 170n32
Guyau, J. M., 21

Hamel, G., 129
Harmony, 62
Hawking, S., 6
Hegel, Georg W. F., 116, 123–124
Heidegger, Martin, 3, 13
Heisenberg, Werner, 146, 161
Heliocentricity, 103, 104–107
Hellwig, B., 55
Helmholtz, Hermann von, 17
Hempel, C. G., 139, 154
Henry of Ghent, 94
Heraclitus, 64–67, 72–73, 77
Herodotus, 77, 131
Hesiod, 55–59
Hesse, Mary, 144–145
Hilbert space, 147, 148, 152, 160
Hippolyte, 60
Hologram, 166n7
Homer, 55
Hultkrantz, A., 46–47

Incommensurability, 63–64
Indefiniteness, 59
Indivisables, 79. See also Atomism
Inertia, 106, 107–108, 136
Infinite divisibility, 3, 62
Infinite velocity, 107, 115–116
Innate structures, 13
Inquisition, 100, 104, 105
Insane ideas, 2
Insanity, worship of, 173n91. See also Psychopathology
Instincts, 13, 106
Intelligence, levels of, 20–23. See also Cognition; Reason
Isotachys, 77–78, 131. See also Motion

James, William, 24
Jaspers, Karl, 33
Judaism, 91
Jung, Carl, 171n55

Kala, 1
Kant, Immanuel
 antinomies of, 122–123
 and geometry, 126–127
 spatiotemporal conception of, 116, 118–123
Kantianism, 17
Kashina, T. I., 45
Kekinema, 77, 78. See also Atomism
Kenon, 85–86, 108. See also Democritus's void
Kepler, Johannes, 104, 106, 130
Kessidi, F. Kh., 57, 66, 167n17
Kierkegaard, Sören, 47
Kinematics, 128–129, 137
Kinesthesia, 15–16
Klein, Felix, 128, 131
Knowledge, sensory vs. rational, 75
Komarova, V. Ya., 77
Konopolev, N. P., 133
Korchazhinskaya, V. I., 26
Koyré, Alexandre, 95
Kronos, 56, 177n7
Kuznetsov, B. G., 86, 107, 113, 115

Lagrange, Joseph Louis, 9, 130, 131
Lakatos, Imre, 89, 125
Language
 and child's concepts, 22, 23–24
 and Euclidean geometry, 164
 Heraclitus on, 65–66
 observational, 162–163
 ordinary, 164
 and spatial perception, 16, 20–21
Lateralization, 28, 41
Leach, E., 42
Lec, Jerzy, 7
Leibniz, Gottfried Wilhelm von, 8, 112–113, 116–118, 186n188
Length, 139, 144, 145, 162
Lenin, V. I.
 on analysis of matter, 151
 on objective reality, 14
 on science and mythology, 2
 on time, 48
Leroi-Gourhan, A., 31
Leucippus, 73–74, 79
Lévi-Strauss, C., 30, 166n5, 171n58
Lévy-Bruhl, L., 3, 30, 42
Light, velocity of, as finite, 141
Light ray, and physical theory, 140, 141, 143
Lindsay, Peter, 22
Linguistic relativity, 22–23,

193n100. *See also* Sapir-Whorf hypothesis
Logical positivism, 127, 139
Logos, 62, 65
Lorentz, H. A., 9, 137, 141
Losev, A. F., 55–56, 75, 80, 115–116
Losev, A. V., 7
Lotman, A. Yu., 44
Luria, A. P., 32
Lur'ye, S. Ya., 74, 75, 77

Magic, 46
Magnitudes, fundamental. *See* Basic concepts
Mahler, Gustav, 3
Makovel'skiy, A. D., 76
Mandel'shtam, L. I., 141
Maori, tribal space of, 50
Marx, Karl, 31
Marxist-Leninist philosophy, 31–32
Mass, as concept, 129
Materialism. *See* Dialectical materialism; Greek philosophy, early
Materialism and Empirocriticism (Lenin), 10
Mathematical Principles of Natural Philosophy (Newton), 8–9
Mathematics
  and language, 66
  and physics, 126–127
Matisse, Henri, 3
Matter, 3, 62, 74, 108, 112. *See also* Amera; Space, as substantial
Maxwell, James C., 9, 141
Measurement, of time
  and Aristotle, 83–84, 182n103
  and Bruno, 105
  and Copernicus, 104
  and Newton, 114
  and Plotinus, 90
  and St. Augustine, 96, 98–99
Mechanics. *See also* Classical mechanics; Newton, Sir Isaac
  of Galileo, 107
  and isotachys, 78
  and optics, 140–145
  theoretical vs. empirical levels of, 130
Medieval philosophy, 99, 183n128
Meditations on First Philosophy (Descartes), 109–110
Meletinskiy, Ye. M., 36, 41
Melissus, 68

Memory, 97
Mental processes. *See* Cognition; Perception
Metamythology, 166n5
Metaphysics (Aristotle), 94
Metatheoretical regulator, 10, 160–162
Meyerson, Émile, 30, 32, 115
Milesian school, 38, 57, 58–62, 177n17, 178n22
Mill, John Stuart, 138
Mind, 11–12
Minkowski, Hermann, 143
Molchanov, Yu. B.
  on Copernicus, 104
  on Heraclitus's *Logos*, 65
  on time, 70, 83–84, 90, 97, 99
Monads, 116–117
Monotheism, 52–53
Mostepanenko, A. M., 143
Motility, 15
Motion
  and Aristotle, 82–83
  and atomism, 72–74, 77–79
  and axiomatics, 128–129
  and Descartes, 111
  and Hegel, 124
  and Heraclitus, 66–67
  and Kepler's laws, 106
  Newtonian laws of, 78, 110–111
  Sextus Empiricus on, 78
  and Zeno, 72–73
Mythology
  contradictions in, 42–43
  dynamics in, 34–36
  Greek, 34, 35, 38, 55
  and *kamlaniye* ritual, 36, 38–39, 41–42
  and multilayered space, 36–38
  and pathopsychology, 41–42
  primacy of water in, 34
  and relativity, 39–40
  and religion, 46–48, 96–97
  and time, 39, 40–42, 44–45, 46–48, 96–97

Nalimov, V. V., 153
Nativism, 17
Natural laws, 98, 132
Natural philosophy, 2, 4, 6–7
The Natural Philosophy of Time (Whitrow), 11–12
Necessity, 60–62
Newton, Sir Isaac. *See also* Classical mechanics

Newton, Sir Isaac (cont.)
  and absolute time, 9, 68, 80, 104, 114–116
  Frank on, 136
  and geometry, 129, 133–134
  and laws of motion, 78, 110–112
  and reconstruction, 130–134
  and space, 57, 112–114, 116
  and wave theory, 9, 140
  and Xenophanes' One, 68
Nicholas of Cusa, 53, 93
Nomos, 62
Nonclassical theories, modern, 148, 149, 150–151
Norman, Donald, 22
Nothingness. *See* Absolute space
Noumena, 121
Numbers, and space, 62–64

Objective idealism, 1
Observational verification, 147. *See also* Empirical structures
Omnipresence, divine, 93, 94, 96–97
One, the, 67–70
On the Revolution of the Heavenly Spheres (Copernicus), 102
Ontogeny, and phylogeny, 5, 17, 168n17
Operational level, 140, 141, 162
Optics, 140, 141
Oracle, 61, 62
Order, oases of, 36–38
Oriental philosophy, 91
Osiander, Andreas, 102

Paleopsychology, 12
Pantheism, 53
Parmenides, 63, 64, 69–70
Particles, 10, 161
Past, the, 41, 96, 97–98, 183n133
Patrizi, Francesco, 101–102
Peiper, Albert, 18, 168n17
Penrose, Roger, 133–134
Perception
  in children, 18–20, 170n35
  and empirical objects, 139
  innate structures of, 13, 17 (*see also* Primary structures)
  microgenesis, of, 169n27
  primitive levels of, 37
  and relativistic physics, 5, 137–138, 147
  spatial, 5, 15–16, 18, 19–21, 106
  and theory, 4–6

Peripatetic doctrine, 96, 100
Perspective, in painting, 169n25, 184n140
Phenomenology, 13–14
Philolaus, 63, 79
Philosophy of Nature, The (Hegel), 124
Philosophy of Physics (Bunge), 154
Phylogeny, and ontogeny, 5, 17, 168n17, 171n55
Physical theory
  axiomatization of, 152–165
  classical, 8–9, 103 (*see also* Classical mechanics)
  and empirical objects, 139
  and geometry, 108, 125–129
Physical theory, modern. *See also* Quantum theory; Relativity, special theory of
  and Aristotelian concepts, 131
  and empirical base, 144–145, 147
  empirical vs. theoretical in, 136
  and Hesiod's chaos, 57–58
  and Platonic doctrine, 79–80
  and psychology, 4, 165
Physics (Aristotle), 94
Piaget, Jean, 41, 43
Planck, Max, 9, 159–160
Plato, 7, 79–81, 181n96
Plotinus, 90, 95
Plurality, 70–71, 73
Podgoretskiy, M. I., 10, 158
Poincaré, Jules Henri, 137, 141, 148
Points
  in Leibniz, 116–117
  in Pythagoreanism, 62, 63–64
Polytheism, 52
Popova, L. T., 26
Popper, Karl, 89
Post-Aristotelian philosophy, 90–91
Posterior Analytics (Aristotle), 125–126
Prediction, and spatiotemporal continuum, 14–15
Prelogicality, 42–44
Present, the, 61, 70, 96
Pribram, Karl, 2
Primary structures, 13–14
Primitive thought
  components of, 42–44
  and modern viewpoint, 32–33
  and preconceptual intelligence, 21
  and relativity, 40
  spatiotemporal concepts in, 19, 35, 37, 38–42

Principia (Newton), 126, 135
Principles of Philosophy (Descartes), 109
Prokof'yeva, Ye. D., 37
Protophysics, 155–158, 162
Pseudoaxiomatization, and physical theory, 152–153
Pseudo-Dionysius, 95, 183n131
Psychology. *See also* Perception
 developmental, 12, 21–22, 41
 historical, 30
 and physics, 4, 165
Psychopathology, 12, 24–30, 41
Ptolemy, 92, 105. *See also* Cosmos, Aristotelian-Ptolemaic
Pythagoreans, 62, 63, 65, 73, 79

Quantum theory, 9, 10
 and classical mechanics, 158–159
 E-structure of, 146, 147
 Finkelstein on, 150
 and Hilbert space, 160
 and nonclassical magnitudes, 150
 and spatiotemporal order, 148
 universality of, 192n88
Quarks, and Greek atomism, 74
Quine, W., 127

Ramnoux, Claude, 67
Reason, 30. *See also* Cognition; Intelligence, levels of
 and early Greek philosophy, 56, 58
 opinionative, 69–70, 171n52
Rectilinear inertia, 106–107, 110
Reducibility algorithms, 126
Reichenbach, Hans, 118, 119, 121
Reincarnation, and cyclic time, 42
Relational concepts, 141–142
 and Heraclitus, 65
 and Leibniz, 116–118
 and relativity, 142
 and substantial concepts, 118, 121, 142, 188n21
Relative space, 112, 116, 137
Relative time, 80, 112, 116, 137
Relativity, special theory of
 and classical mechanics, 116, 134, 143
 and flow of time, 96–97
 interconnection of zero and infinity in, 116
 and mythological time, 39–40
 relational concept in, 142–143
 substantial concept in, 141–142
 T-structure in, 142–143
 universe vs. microworld in, 143–144
Religion. *See also* Christian theology; Judaism
 and mythology, 46–48, 96–97
 and scientific thought, 48
 and spatial conception, 48–54
 and temporal conception, 46–48, 96–97
Renaissance, 99–108
Renovation, 77, 78
Riemann, B., 140
Riemannian space, 148, 152, 161
Rigid body
 and classical mechanics, 158, 159
 and physical theory, 140, 141, 143
Ritter, Heinrich, 6–7
Rozhanskiy, I. D., 57
Rumer, Yu. B., 149–150
Rushd, Ibn, 91
Russell, Bertrand, 91, 117, 121, 139
Russian epics, temporal inversion in, 183n133
Rybakov, B. A., 45
Rybalko, Ye. F., 15, 18, 29

Saint Augustine, 95, 96–98, 183n131
Sapir-Whorf hypothesis, 22–23, 165. *See also* Linguistic relativity
Schliemann, Heinrich, 38, 173n83
Science
 and Kantian cosmology, 118–119
 and mythology, 2
 and phenomenology, 13–14
Sechenov, I. M., 21
Sensorium Dei, 1, 95, 98, 118, 120
Sextus Empiricus, 78, 89–90
Shaman. *See* Mythology, and *kamlaniye* ritual
Shreider, Yu., 3
Shvyrev, V. S., 188n23
Simplicity, and relativity, 149
Simplicius, 68, 80
Simultaneity, 115, 137, 189n42
Smorodinskiy, Ya. A., 10, 158
Social phenomena, and spatial relations, 16, 50–51
Soul, 39, 49, 95–96, 98–99
Space
 empty (*see* Absolute space)
 as eternal, 60

Space (cont.)
  as extensional, 108, 135, 188n21
  as finite, 53
  as infinite, 93, 95, 103, 178n22
  as macroscopic, 149, 151, 190n58
  mathematical shape of, 76, 79–80
  as multilayered, 37
  as place, 85, 101, 117
  as relational, 84–88, 135, 188n21
  sacred, 50
  as substantial, 76, 101–102, 188n21
Space-time
  four-dimensional, 142–143
  of Galileo, 133–134
  and nonclassical physical theory, 149
  symmetry of, 131–132
Spatiotemporal concepts, basic status of, 162–165. *See also* Empirical structures
Spatiotemporal continuum, evolution of, 15
Species differences, 18
Spinoza, B., 53, 109
"Stadium, The" (Zeno), 71, 72
Steblin-Kamenskiy, M. I., 37
Subjective idealism, 1
Subjectivity, 98, 121
Substantiality, 76, 101–102, 118, 142, 188n21
Sun, and time measurement, 182n103
Superspace, 162

Telesio, Bernardino, 101, 104
Temporal duration. *See* Duration
Temporal orientation, 47, 174n104
Thales, 57, 61. *See also* Milesian school
Theogony (Hesiod), 7, 55–59, 60
Theophrastus, 69
Theoretical-empirical dualism, 144–145, 190n48
Theoretical structures, 138–152, 163–164
Thom, René, 164
Thompson, George, 35–36, 57
Three-dimensional space, as inherent notion, 13
Timaeus (Plato), 81
Time
  as cyclic, 39, 41–42, 44–45, 60, 97
  as dynamic, 61, 64
  as eternity, 60, 90, 114
  flow of, 96–97, 99
  as linear, 41, 45, 97
  as macroscopic, 149, 151, 190n58
  and motion, 82–84, 128, 179n52
  as numerical, 81–82
  spiral model of, 45
Timelesness, vs. infinite time, 64–65, 99
Topophilia, 49
Topos, 85–86, 108
Totemism, multilayered world of, 35–36. *See also* Mythology
Tribal space, 49–51, 172n76
T-structures. *See* Theoretical structures

Unified theory, 142, 160–162

Variable, concept of, 112, 129–130
Vavilov, S. I., 11
Vilenkin, N. Ya., 3
Vision 15, 17, 97, 169n24
Void. *See* Absolute space; Democritus's void
Vygotskiy, L. S., 30, 32

Water, as fundamental principle, 34
Wave optics, 141
Wave theory of light, 9
Wheeler, J. A., 58, 130, 142
Whitrow, G. J.
  on Leibniz, 118
  on time, 11–12, 110, 114
  on Zeno's paradoxes, 71, 72
Wigner, E. P., 132
Wilkerson, T. E., 121

Xenocrates, 79
Xenophanes, 64, 67–69. *See also* Eleatic school

Yanovskaya, S. A., 126
Yilmaz, Hüseyin, 167n16

Zadeh, Lotfi, 3
Zederblom, N., 46
Zen, and subatomic physics, 3
Zeno, 64, 70–73, 78, 124, 180n60
Zoroastrianism, 1
Zotoy, A. F., 143
Zurvan, 1

A113 0679429 5

BD 632 .A4513 1986

Akhundov, Murad Davudovich.

Conceptions of space and
 time--sources, evolution